The Technology, Strategy, and Politics of SDI

About the Book and Editor

The Strategic Defense Initiative (SDI) to develop a ballistic missile defense (BMD) system has both short-range and long-range risks as well as potential benefits. For the most part, however, strategic, technological, and political issues relevant to SDI have been analyzed in isolation from one another. This book provides a more inclusive framework for assessing the possible development and deployment of a BMD system by the United States or the Soviet Union. Contributors discuss the risks for arms race stability, probable reactions of the Soviet Union to any U.S. space-based defense system, and implications for the stability of extended deterrence commitments to NATO European allies. They also evaluate Soviet research and development programs in missile defense that must be considered in any extrapolation of the requirements for U.S. deterrence in the next several decades.

Stephen J. Cimbala is professor of political science at Pennsylvania State University.

The Technology, Strategy, and Politics of SDI

edited by
Stephen J. Cimbala

Routledge
Taylor & Francis Group
LONDON AND NEW YORK

First published 1987 by Westview Press

Published 2019 by Routledge
52 Vanderbilt Avenue, New York, NY 10017
2 Park Square, Milton Park, Abingdon, Oxon OX14 4RN

Routledge is an imprint of the Taylor & Francis Group, an informa business

Library of Congress Cataloging in Publication Data
The Technology, strategy, and politics of SDI.
(Westview special studies in national security and defense policy)
1. Strategic Defense Initiative. I. Cimbala, Stephen J. II. Series.
UG743.T43 1987 358'.1754 86-28096

ISBN 13: 978-0-367-29647-6 (hbk)
ISBN 13: 978-0-367-31193-3 (pbk)

To my wife
and
to my mother

Contents

PART FOUR
Soviet Strategic Policy and the Impact of SDI

PART FIVE
SDI and Arms Control: Compatible or Conflictual?

Acknowledgments

The editor gratefully acknowledges the assistance of Barbara Ellington of Westview Press and Diane Wolf of Pennsylvania State University, Delaware County Campus, in preparation of this manuscript for publication. Arguments by the contributors represent their personal work and not the views of the U.S. government or other employing agencies.

Stephen J. Cimbala

Acronyms

AD	assured destruction
ALCM	air-launched cruise missile
ASAT	antisatellite
ASW	antisubmarine warfare
ATBM	antitactical ballistic missile
	antitheater ballistic missile
AWAC	Airborne Warning and Control System
BMD	ballistic missile defense
C^2	command and control
C^3	command, control, and communications
C^3I	command, control, communications, and intelligence
CIA	Central Intelligence Agency
DE	damage expectancy
DPB	defense-protected build-down
ECM	electronic countermeasures
EMP	electromagnetic pulse
FOBS	fractional orbital bombardment system
GLCM	ground-launched cruise missile
HOE	Homing Overlay Experiment
ICBM	intercontinental ballistic missile
INF	intermediate-range nuclear forces
IRBM	intermediate-range ballistic missile
kt	kiloton
LCC	launch control center

LNO	limited nuclear option
LRINF	Long-range Intermediate Nuclear Forces
LWIR	long wavelength infrared
MAD	mutual assured destruction
MARV	maneuverable reentry vehicle
MAS	mutual assured survival
MIRV	multiple, independently targeted, reentry vehicle
MT	megaton
NASA	National Aeronautics and Space Administration
NATO	North Atlantic Treaty Organization
NCA	National Command Authority
NPG	Nuclear Planning Group
NUWEP	Nuclear Weapons Employment Policy
OTH	over the horizon
PBV	postboost vehicle
penaid	penetration aid
PSI	pounds per square inch
R&D	research and development
RSD	realistic self-deterrence
RV	reentry vehicle
SAC	Strategic Air Command
SALT	Strategic Arms Limitation Talks
SAM	surface-to-air missile
SDI	Strategic Defense Initiative
SDIO	Strategic Defense Initiative Organization
SIOP	Single Integrated Operational Plan
SIR	shuttle imaging radar
SLBM	sea-launched ballistic missile submarine-launched ballistic missile
SNDV	strategic nuclear delivery vehicle
SRBM	short-range ballistic missile
SSBN	ballistic missile submarine
SSN	nuclear-powered attack submarine
START	Strategic Arms Reduction Talks
TERCOM	terrain contour mapping
TVD	theater of military operations

Introduction: Strategic Defense—Ever Promising, Ever Frustrating

Stephen J. Cimbala

The U.S. dialogue on strategic defense includes disputes about matters as basic as whether it is desirable to have defenses at all. As the chapters in this book indicate, the debate about active defenses versus ballistic missile threats continues to reflect the most fundamental disagreements. Even those who think ballistic missile defense (BMD) desirable or feasible will not concur about the "how" and "when" of its deployment.

The arguments about BMD can be divided into theological and pragmatic categories. Theological disputations reflect ongoing scholarly concern about whether deterrence by "denial" or by "punishment" is more effective. In academia this debate is very polarized. One faction holds that deterrence by punishment is the only adequate basis for stable strategic nuclear deterrence between the superpowers. Another contends that mutual assured destruction (MAD) or assured retaliation is no longer credible in the face of new technologies and an ambitious Soviet challenge.

The Reagan administration has departed from precedent by taking an explicitly heterodox theological position; the Nixon, Ford, and Carter administrations kept the research and development lines warm but avoided unambiguous commitment to a preferred theory of deterrence. President Ronald Reagan struck preemptively against his own defense establishment in his March 23, 1983, speech, which called for defenses that would make nuclear weapons obsolete. Experts in the administration and outside of it are still attempting to explain how the president came to make this statement.

Within the administration, the Strategic Defense Initiative Organization (SDIO) has been organized to turn the president's vision into something real, or at least realistic. The pragmatic issues concerning

BMD deployment have thus come to the fore, and some of those issues are discussed in this book. There are technical issues regarding what systems can be built and where they can be deployed, if at all. Both terrestrial and space-based deployments are contemplated by contractors and SDIO officials, and thus both must address the contemplated by contractors and SDIO officials, and thus both must address the issue of how to "defend the defense" if it is based in space. This question of defending the defense may be more formidable than deploying space-based defenses in the first place. Nor is the successful weaponizing of space defenses, or of the defenses for space defenses, necessarily the most formidable challenge. The truly demanding parts of SDI may appear in the development of the corequisite command, control, and communications (C^3) and battle management software. Much controversy already has erupted within the administration about this last point, with computer experts numbered both among the optimists and the pessimists.

The political implications of SDI deployment, especially in regard to U.S.-Soviet relations and NATO alliance cohesion, need to be elaborated in any consideration of the effect of the SDI upon deterrence stability. Soviet declaratory policy consistently has opposed U.S. deployment of "space strike" weapons, which is seen as compatible with a U.S. capability for strategic preemption. However, the USSR perseveres in research and development, not to mention deployment of an existing system, that might provide a basis for a future SDI of its own. Some might question whether the U.S. program should even be called an "initiative" instead of a "response" to consistent Soviet efforts in BMD, air defense, and civil defense on a scale not matched in the United States or in NATO Europe.

Thus, the U.S. problem in deploying BMD, with regard to U.S.-Soviet relations, is twofold. Washington must manage the competition with the Kremlin so as to preclude a Soviet "breakout" from the Antiballistic Missile (ABM) Treaty that might leave the West in a comparatively inferior position. Washington also must avoid the appearance of deploying a U.S. BMD that would jeopardize Soviet second-strike capabilities, which are thought to be essential for stable deterrence by mainstream U.S. strategists. This is a difficult enough challenge even without allies. Bringing them along poses additional problems.

First, it is not clear that the European allies of the United States will react in any consistent way to the proposal for BMD deployment. Their reactions are to some extent "scenario dependent" upon their assumptions about what the Soviets will do. If Europeans assume that the USSR will react to any U.S. BMD deployment with a BMD deployment of its own, then they are likely to be pessimistic about the implications of BMD for deterrence stability. This is especially the case for the British and French; even partial Soviet BMD could threaten the viability of their independent national deterrents. On the other hand, if Europeans perceive the USSR to be intent upon deploying its own defenses no matter what the United States does, then they may see U.S. deployments as inevitable and want to share in a "two-way street" of defense contracts. European reactions also will be conditioned by expectations about what kind of system the United States or the USSR might deploy. Deployment of a U.S. BMD to protect missile silos and other counterforce targets wold have implications quite different from deployment of BMD to protect cities, assuming the latter is at all feasible.

Second, an issue of more immediate pertinence to Europeans will be the effects of missile defenses on the balance of theater conventional and nuclear forces. Here the "seamless web" of deterrence becomes quite murky. NATO antitheater ballistic missile (ATBM) defenses based in Europe might be useful in protecting key military and command targets against Soviet preemption. However, such deployments by NATO could be met by a Soviet ATBM, which would nullify much of the highly touted "deep attack" conventional force improvements to which NATO seems committed. The effort to raise the nuclear threshold through the deployment of enhanced technology conventional weapons systems might be negated by very competent Warsaw Pact theater missile and air defenses. Conversely, if NATO does not deploy an ATB but the USSR does, then the Soviets move closer to a position of uncontested "escalation dominance" in the event that crises spill over into war in Europe. Already vulnerable to equivalent Soviet strategic offensive forces and superior theater nuclear forces, NATO cannot ignore the potential force multiplier provided by a Soviet ATBM unmatched by NATO offensive countermeasures and/or a Western ATBM deployed in Europe.

As if these are not enough, there is the added spice of controversy within the legislatures and bodies politic on both sides of the Atlantic, for NATO is an alliance of democratic and disputatious bedfellows. The U.S. policy debate regarding defense deployments will be especially acrimonious as the implications of Gramm-Rudman, tax reform, and Congressional skepticism about defense and social programs become clear. Within the defense budget, the struggle between strategic nuclear and other programs also will affect that policy debate. The U.S. SDI program runs the risk of being hostage to the latest congressional appropriations hearing or press exposé. The Reagan defense program from fiscal year 1981 to 1985 already has built up a momentum that will be expressed in outlays during the latter part of the decade, which leave less room for maneuver in the short run. There is also the inevitable skepticism of the U.S. public in general about the entire point of missile defenses, given that the public has been fed nuclear war scenarios that are invariably senseless holocausts. The political survivability of defenses is more problematical than any of the cases made for them on technical or strategic grounds.

Contributors' Perspectives

Gary Guertner considers the relationship between strategic offenses and defenses within the context of the Reagan administration's new definition of defense. The Reagan proposals must be considered in light of more than a decade's evolution in the U.S. doctrine of strategic offense. Strategic offensive forces in fact are not intended to be blunt instruments of revengeful, massive societal destruction. During the past two decades efforts have been made on many fronts to increase the range of targeting options available in the Single Integrated Operational Plan (SIOP) for employment of strategic nuclear forces. Nuclear weapons employment policies signed in 1974 divided targets into four principal groups, and it is important to note that none of these groups included civilians as targets per se. In response to this policy, targeting plans were adjusted to give even more priority to military targets. In a sense this brought U.S. employment policy into line with the earliest declaratory policies of the Kennedy administration, although they were later abandoned. The Carter administration further implemented the Schlesinger doctrine of 1974 by deemphasizing economic targets and emphasizing the importance of survivable forces and C^3 for fighting a limited nuclear war. The Reagan strategic modernization program continues the Schlesinger limited nuclear war. The Reagan strategic concept, however, which calls for very competent defenses and greatly reduced offenses, unintentionally might reverse the progressive development of more selective options for controlled warfighting. It might do this by protecting strategic forces and command centers in the near term, reducing the attractiveness of limited counterforce options, and making urban-industrial areas more subject to destruction, even with limited offenses.

This paradoxical aspect of defense deployments also is explored by Donald Snow. The most basic issue, in his judgment, is whether missile defenses reinforce or detract from deterrence. Obviously this issue is not yet decided. Much of the policy debate is rooted in both orthodox and heterodox positions that have their roots in formulations going back to the 1940s. The orthodox position—mutual assured destruction—is endorsed either as a preferred policy or as an inescapable fact and regards population defenses as destabilizing and improbable. Partial defenses of retaliatory forces are considered justifiable by some MAD proponents but are considered of dubious benefit by others. The heterodox position argues that the objective of deterrence is not to prevent all-out nuclear attacks but to deter the limited hostilities that are much more probable. U.S. defense deployments would complicate Soviet attack scenarios even if the defenses were not perfect. Snow argues for a third position—realistic self-deterrence. This position assumes that the first foreign policy priority of both superpowers is avoidance of nuclear war with the other. Thus, each side deters itself; the secret of deterrence strategy is how to maintain that condition. The defense thus will involve a delicate balance between maintaining the basis of deterrence stability through offensive retaliation and amending that condition through cooperation with the USSR, including arms control agreements that reduce the anxieties of both sides.

Leon Sloss and Seymour Weiss object to the either/or tenor of public debate about the SDI. Future U.S. deterrence requirements, in their judgment, will include a balanced strategic posture with both offenses and defenses. Many critics of the SDI are more concerned about preserving the ABM Treaty than they are about exploring the prospects for missile defense; other proponents of the SDI are oblivious to arms control. Sloss and Weiss agree with SDI critics that a "perfect" defense is not within reach in the foreseeable future, but they note that the Soviet interest in BMD cannot be dismissed easily. If MAD is an inevitable condition, then the USSR has wasted its considerable investments in BMD and air defenses. Although the MAD perspective needs revision, the "techtopian" alternative of mutual assured survival may be fatally optimistic. According to Sloss and Weiss, both MAD and the techtopian mutual assured survival perspectives are mistaken. Reliance on nuclear weapons cannot be invented away, for strategic or theater war. U.S. strategy for more than a decade has evolved in the direction of more realistic and selection options for the use of offensive forces. Defenses can contribute to a strategy of selective, limited use of nuclear options by adding uncertainty to the calculations of an attack planner. The near-term contributions of defenses (rather than deliberations about "perfect" defenses) should have pride of place on the policy agenda. One might in fact pursue research and development toward long-term goals of improved defenses while deploying systems of limited effectiveness. Such systems might improve stability if they reduced the temptation for first strikes against strategic forces and command targets.

The questions raised by Sloss and Weiss are very pertinent to the cases against and for point defenses made by Jonathan Stein and David Sorenson. Arguing against such hard-point defense of U.S. intercontinental ballistic missile (ICBM) silos, Stein contends that efforts to revise or debunk mutual assured destruction are misconceived. The plausible case has not been made, in his judgment, for counterforce warfighting or deployments of offenses and defenses pursuant to that end. Nor are the critics of MAD accurate in their skepticism about the results of the Strategic Arms Limitation Talks (SALT) process, especially in their appraisal of SALT I. Ironically, according to Stein, advocates of defenses now contend that they can bring about the de-MIRVing of Soviet offensive forces, which could have been accomplished in a more straightforward way through arms agreements specific to that point. Point defenses are of a piece with offensive warfighting strategies in their misbegotten deterrence logic, according to Stein. Under very circumscribed conditions, including vastly reduced offenses, point defenses might contribute to stable deterrence. Those conditions are more theoretical than politically realistic. Each side will expand defenses in reaction to the other's offenses because there are only two stable equilibrium points in the nuclear era: mutual assured destruction and mutual assured survival. Points in between are unstable.

David Sorenson is more optimistic about the prospects for point defense. His reading of the deterrence implications of BMD and of the historical record of offense-defense interaction leads to conclusions very different from those of Stein. Offensive and defensive dominance have alternated with historical changes in tactics and strategy. The present balance of U.S. and Soviet counterforce capabilities results in asymmetrical countersilo capabilities favoring the USSR to the detriment of U.S. deterrence. Although it might be argued that no Soviet first strike is without significant risks, these risks might be judged tolerable under crisis conditions of perceived desperation. History provides examples of attacks launched under conditions of desperation and uncertainty, although it also provides counterexamples of military advantages that were not exploited. Although nothing can match the destruction of a major nuclear war, other Carthaginian outcomes are not historically unknown. Because war cannot be precluded as a possibility, point defenses can contribute to stability under certain conditions. Sorenson suggests that point defenses of silos could raise the attack price for the destruction of U.S. ICBMs if the defenses were technically feasible, as they appear to be. Emerging technologies that will provide additional counterforce capabilities and improved accuracies for offensive forces may stimulate further interest in point defenses.

The transatlantic challenge will be as problematical for the SDI as will its negotiation with the Soviet Union and within the U.S. government. David Schwartz addresses the prospective debate about the SDI and strategic defense within the Atlantic alliance. European governments and academic circles have paid little attention to the issue of BMD until recently. During the 1970s the ABM Treaty and the low priority accorded the BMD seemed to have fostered disregard of the issue in favor of others. A strategic world of very competent offenses and nugatory defenses was taken for granted. NATO flexible response strategy rests on the capability for three levels of response to aggression: direct defense, deliberate escalation, and general nuclear response. Thus, U.S. strategic offensive forces are the ultimate guarantee of deterrence credibility. The implications of strategic defense, given this background, are profound. Experts disagree about the contents of those implications. Some argue that the SDI would strengthen extended deterrence. The United States will be more willing to fight on behalf of allies if its homeland is defended, in this view. There is the equally persuasive argument, at least to some Europeans, that an effectively shielded United States would be reluctant to take the risks of involvement in nuclear war in Europe. There is also a third possibility—that both superpowers would be more willing to engage in theater conflict behind what were thought to be effective shields. Generally speaking, Europeans are more amenable to the SDI when it is presented as contributory to, rather than undermining of, deterrence based on offensive retaliation. Protection of military targets in Europe by an ATBM might be a more politically and technically feasible alliance objective than comprehensive population defense.

James Wendt also considers the implications of missile defense for Europe. Of fundamental importance is the issue of British and French responses to Soviet deployments of BMD, which might or might not be responses to U.S. initiatives. In the case of Britain, its leaders have articulated an explicit "second national decision center" rationale for their independent strategic nuclear deterrent. This rationale has subtle implications that are not always noted; for example, the rationale implies that the Soviets need to be convinced of the U.S. nuclear guarantee to Europe more than the Europeans do. The notion of proportionality is central to British definitions of targeting objectives; the USSR must not consider cost beneficial the destruction of Britain in the context of a continuing war with the United States. Expansion and modernization of the British ballistic missile submarine (SSBN) force will influence targeting decisions, as will actual patrolling patterns under day-to-day and crisis conditions. Proportional deterrence is central to the French conception of tasking for strategic nuclear forces. This meant to theorists such as General Pierre Gallois that the nation's strategic nuclear force could be proportional to the value of the stake it was defending. Although this concept remains at the core of French strategic thinking, both have evolved. Tactical nuclear weapons may be used to the enemy's intentions and to warn of the risk of escalation. Other doctrinal formulations implied that France might participate in the "forward battle" in West Germany, under the concept of "enlarged sanctuarization," which extended French national territory to include additional areas coming under its putative protective umbrella. French land- and sea-based strategic forces will be substantially modernized by the mid-1990s. French targeting doctrine seems to have evolved from a countercity employment policy to an "enlarged anti-cities" policy emphasizing destruction of the Soviet economy and industrial recovery potential. A Soviet BMD system that was less than adequate to defeat any U.S. attack still could offer some significant protection against smaller national forces, especially if those forces were striking in retaliation against alerted Soviet defenders. In response, the British and French could move to a more cruise missile–oriented posture, adjust targeting plans (including joint targeting arrangements), emphasize defense suppression, or build additional forces.

Much of the evaluation of the SDI depends upon assumptions about the Soviet view of BMD. Daniel Gouré suggests that the USSR views the U.S. SDI as a significant challenge in two senses. The United States might develop a combined offensive and defensive posture that would preclude Soviet preemptive options. A U.S. SDI also might provide additional U.S. offensive options. Soviet planners might perceive U.S. strategic defenses, coupled with improved NATO theater nuclear forces, as providing more credible extended deterrence. The SDI also presents a scientific-technical challenge to the Soviet Union. It reinforces the Soviet expectation that offenses and defenses coexist in dialectical relationship. The USSR continues to improve its strategic defenses while improving its offensive capabilities qualitatively and quantitatively. The USSR is improving air and ballistic missile defenses already deployed. It also is investigating advanced technologies with particular emphasis upon directed-energy weapons. Thus, Soviet perspectives on the ABM Treaty may be ambivalent. The Soviets might prefer limited competition under the treaty umbrella with room for ambiguous deployments to an all-out competition with the United States that would follow treaty abrogation. Although the USSR has stated its intention to counter U.S. SDI deployments with offensive countermeasures, the Kremlin is uncertain whether this can be accomplished in practice.

Richard Thomas notes that for many years the Soviet Union has been planning for the military uses of space, including the deployment of strategic defenses that might operate there. Intrinsic to Soviet military doctrine is the notion of damage limitation, which can be accomplished by active and passive defenses, together with very competent strategic offensive forces appropriately tasked. Thomas' focus is upon the Soviet interest in active defenses, including traditional air defenses, ballistic missile defense, and space defense. Soviet doctrine has evolved toward ever more sophisticated frameworks for analyzing BMD and military space operations. Military space missions identified as important as early as the 1960s included reconnaissance and launch detection, test monitoring, damage assessment, communications, and navigation. Early concern with battle management/C^3 issues also was apparent in the Soviet literature. The Soviets have not "mirror imaged" U.S. technological approaches for similar missions; they have adapted technology to their expectations about U.S. behavior as understood through Soviet conceptual lenses. The USSR, for its part, has never accepted the doctrine of mutual assured destruction and has acted on the assumption that offensive weapons, however formidable, can be countered. Current Soviet strategic defenses include layered terminal defenses with ground-based radars to activate the defensive missiles and phased-array radars for battle management. In the future surface ship or submarine-based BMD may be used for exoatmospheric intercept. The development of a Soviet military space presence may coincide with a definition of space as a theater of military operations in itself.

The relationship between the SDI and arms control is considered by Keith Payne and Patrick Friel. The proper combination of arms control and BMD deployments, according to Payne and Friel, could lead to a radical, favorable adjustment in the character of superpower strategic deterrence stability. The deployment of BMD for the defense of retaliatory forces only, in order to support deterrence by offensive retaliation, raises problematical issues. The improvement in prelaunch survivability may be offset by the decrement in penetrativity, if the USSR also deploys point defenses. In addition, BMD may not be the most effective way of ensuring prelaunch survivability. Thus, near-term deployments should have the potential for expansion into something more inclusive than silo defenses. One attractive option is a near-term midcourse BMD system that could support a variety of defensive missions using both point and area defenses. This would defend selected assets preferentially. The technology would include space-based active and passive sensors, and a ground-based, infrared homing, nonnuclear impact warhead for defense in the high endoatmospheric or the exoatmospheric altitudes. Two constellations of satellites, one in geosynchronous orbit and a second a low, near-earth orbit, also would be deployed. The second constellation would employ passive long-range infrared sensors or millimeter wave radars to obtain high-quality active and passive images throughout the MIRV deployment sequence, thus allowing tolerable solutions for the problem of discrimination. The sensors also would provide RV impact point prediction. Even very limited arms control agreements would make this near-term, nonnuclear kill BMD system very attractive, according to the authors. They argue that the traditional arms control theory that rejects BMD as destabilizing is misconceived. An arms control regime that reduces offensive forces and permits progressively improved defenses would contribute more to stability in the long run.

Stephen J. Cimbala contends that stability in U.S.-Soviet relations is not a static concept. There is no reason in principle why defenses cannot contribute to stable deterrence. Defense policy choices are choices that ar conditioned by technology, but not determined by it. The idea of "defense dominance" as anything more than a temporary condition is illusory. Defense and offense are interdependent. Whether the defense or the offense dominates the strategic relationship between superpowers may have more to do with their war plans and objectives than it has to do with their technologies. Generally speaking, surprise favors the attacker, but "friction" favors the defender. Thus, it cannot be assumed that strategic defenses will make arms control more likely or deterrence more stable. Defenses of uncertain effectiveness relative to the offenses that oppose them are as likely to cause war as they are to prevent it. This is not apparent in slide rule calculations. The relative value of undestroyed U.S. society exceeds that of a destroyed Soviet society or destroyed Soviet forces under any system of calculation acceptable to the U.S. population. This political equation is more compelling than postattack calibrations. (The realistic possibility of nearly perfect defenses could change this "equation," but not necessarily in a favorable direction. Even nearly perfect defenses could invite suppression against themselves, and the payoff is higher the more effective the defenses are thought to be.) Despite certain cautions, defenses can be contributory to arms control, if we acknowledge that the next century will not be one of celibacy in strategic defense and seek convergence between arms control objectives and denial capabilities.

PART ONE

Are Ballistic Missile Defenses Necessary? Three Perspectives

1

Nuclear War in a Defense-dominant World

Gary L. Guertner

The Strategic Defense Initiative (SDI) debate has revisited many arguments from an earlier period in the evolution of U.S. strategic doctrine. The new debate finds some SDI advocates repeating many of the same arguments against mutual assured destruction (MAD) made during the Nixon years by advocates of limited nuclear war and flexible response. Nixon administration spokesmen also argued that the U.S. president needed options in addition to retaliation against Soviet cities, especially if Soviet reserve forces could retaliate against previously spared U.S. cities.

The Reagan administration often has misrepresented current U.S. offensive doctrine by contrasting its views of a defense-dominant world with the city-busting strategies of the 1960s. Secretary of Defense Caspar Weinberger, for example, has explained that "true believers in the disproven MAD concept hold that the prime, if not the only, objective of the strategic nuclear forces of both the United States and the Soviet Union is the ability to destroy each other's cities."[1]

The president also has argued repeatedly that strategic defense is both militarily and morally necessary. In a White House booklet released on January 1985, he stated that "certainly, there should be a better way to strengthen peace and stability, a way to move away from a future that *relies so heavily on the prospect of rapid and massive nuclear retaliation* and toward greater reliance on defensive systems which threaten no one [emphasis added].[2]

The rhetorical excesses ignore more than a decade of offensive doctrinal evolution and detract from the in-depth assessments required of so major a shift in U.S. strategic posture. It is essential, therefore, to compare current offensive doctrine and future defensive objectives as accurately as possible.

For the past two decades, there has been a continual official effort to increase the range of strategic nuclear targeting options available to the president in a crisis. Alternatives to mutual assured destruction have been developed in the documents, strategies, and force structures of every administration since 1970, including the Reagan administration. Not since the Kennedy administration has a president been confronted with a choice between no nuclear response or the massive unleashing of U.S. strategic forces. These alternatives have been characterized by plans that concentrate against military targets through limited and selective attack options that, in theory, make it possible to control escalation short of attacking cities, to bargain with the Soviets during a nuclear war, and to terminate nuclear conflict at the earliest possible time.

The first official public discussions of these issues came in President Richard Nixon's foreign policy message to Congress on 18 February 1970.

> Should a President, in the event of a nuclear attack, be left with the single option of ordering the mass destruction of enemy civilians, in the face of the certainty that it would be followed by the mass slaughter of Americans? Should the concept of assured destruction be narrowly defined and should it be the only measure of our ability to deter the variety of threats we may face?[3]

A series of studies and directives followed that provided political guidance on structuring more flexible preplanned nuclear responses in the U.S. war plan, or SIOP (Single Integrated Operational Plan). Secretary of Defense James Schlesinger publicly announced the change in targeting strategy. Assured destruction and the old policy of initiating a suicidal strike against the cities of the other side "were no longer adequate for deterrence." He therefore would implement a set of selective options against different sets of targets on a much more limited, flexible scale.[4]

The Nuclear Weapons Employment Policy (NUWEP) instituted by Schlesinger in 1974 set forth the planning assumptions, attack options, targeting objectives, and predicted damage levels needed to satisfy the political guidance developed by the administration. Targets were divided into four principal groups:

- Soviet nuclear forces
- Soviet conventional military forces
- Military and political leadership targets (e.g., command posts)
- Economic and industrial targets (including transportation and energy)[5]

In response to this policy, changes were made in the SIOP that further divided these four groups into specific categories and offered "packages" of strike options that could single out or combine various target categories within the four general groups.[6] Only two of these categories—leadership and economic targets—are associated with mutual assured destruction, and many of those (dams, rail junctions, leadership bunkers) are located outside major population centers. Military targets were given top priority. By adopting the strategy of limited nuclear options, planners reasoned, escalation might be averted short of attacking target categories in major urban-industrial centers.

The Carter administration refined the limited nuclear war strategy by deemphasizing Soviet economic targets (moving still further away from MAD) and stressing the importance of survivable strategic forces and command, control, and communications (C^3) systems required to execute a limited nuclear war.[7]

Subsequently, the Reagan administration produced the Nuclear Weapons Employment and Acquisition Master Plan, which maintained the legacy of limited nuclear warfare and stressed the requirements for strategic modernization including survivable forces and command, control, communications, and intelligence (C^3I) systems to execute selective attack options.[8] In fact, considerable controversy during the administration's first term focused on public discussions of fighting and "winning" limited nuclear wars.

A strategy for controlling nuclear war short of mass destruction may be a false hope, as critics claim. But there is a world of difference between war plans that deliberately (as in the 1950s) provide no options other than surrender or holocaust and those developed throughout the 1970s and early 1980s that at least attempt to mitigate the consequences of nuclear war if deterrence fails. The distinctions between MAD and limited nuclear war have been debated for more than two decades. That debate and whatever wisdom it may have produced should not be ignored as this and subsequent administrations move toward a defense-dominant world that may not provide more security than did its predecessors. If we look to the past, we see that nuclear war was planned in the 1950s on the basis of what U.S. bombers could find—Soviet cities. In the 1990s, the United States may plan war on the basis of what its weapons can hit—again, cities. Like the British and French with limited nuclear resources, U.S. and Soviet planners may be driven toward countercity targeting. We shall then have come full circle at great expense and, in the end, succeeded only in making the world safe for MAD.

The actual conduct of nuclear war could be considerably different from that suggested by the declaratory policies of either the Soviet Union or the United States. Strategic orthodoxy could easily give way to ad hoc strategies based on last-minute military and political judgments or resulting from the chaos caused by a disrupted national command authority. Escalation, collateral damage, and the delayed effects of nuclear weapons (radiation and societal disruption) could drive casualties quickly to "unacceptable" levels or bring about unforeseen consequences even if cities were not attacked directly. Given these possibilities, U.S. leaders and planners need to consider all plausible contingencies and scenarios carefully and think through the impact of strategic defense on deterrence, crisis management, and offensive targeting now rather than in the midst of a future Soviet-U.S. crisis.

The New Strategic Concept

The administration's optimistic predictions for strategic defense often have misrepresented current U.S. strategic doctrine by ignoring the evolution away from mutual assured destruction during the past decade. The "new strategic concept" or vision of the future links the Strategic Defense Initiative to long-range offensive arms reductions. Its goal is to make deep cuts in offensive weapons with the development of strategic defenses during a long, carefully phased transition period. During the next ten years, the United States will seek a radical reduction in offensive nuclear arms, followed by a period of mutual transition to effective nonnuclear defense forces as technology makes such options available. In a final "ultimate period," strategic defenses may make it possible to eliminate all nuclear weapons.

Paul H. Nitze, chief negotiator at the U.S.-Soviet arms control talks and reportedly the author of the new concept, described the three envisioned phases in detail during testimony before the Senate Foreign Relations Committee.

1. The near term: For the near term (through the mid-1990s), the United States will continue to base deterrence on the ultimate threat of nuclear retaliation. Today's technology provides no alternative. That being said, the United States will press for radical cuts in the number and power of strategic and intermediate-range nuclear arms.

2. The transition period: Should a transition be possible, arms control would play an important role. The United States, for example, would seek continued reductions in offensive nuclear arms. Concurrently, it envisages the beginning, on both sides, of testing, developing, and deploying survivable and cost-effective defenses, with particular emphasis on nonnuclear defenses. Deterrence thus would begin to rely more on a mix of offensive nuclear and defensive systems, instead of on the threat of offensive nuclear arms alone. The transition would continue for some time, perhaps decades.

3. The ultimate period: Given the right technical and political conditions, the United States would hope to continue the reduction of all nuclear weapons down to zero. The total elimination of nuclear weapons would be accompanied by widespread deployments of effective nonnuclear defenses. Were the United States to reach the ultimate phase, deterrence would be based on the ability of the defense to deny success to a potential aggressor's attack—whether nuclear or conventional. The strategic relationship could then be characterized as one of mutual assured security.[9]

Assuming that the Soviets could be persuaded to cooperate in the transition to a defense-dominant world (a position they now publicly reject), it is important not to lose sight of the continued, long-term role of offensive weapons. During the near-term phase, for example, deterrence would continue to be based on the threat of nuclear retaliation. Offensive modernization programs would continue even if arms control agreements were to succeed in driving down total force levels.

The transition period calls for a mix of offensive and defensive systems that could be maintained (and modernized) for decades. Nuclear weapons, offensive strategies, and targeting policies would be required well into the next century. It is essential, therefore, that strategic planners carefully assess the probable impacts of such strategic shifts on U.S. and Soviet targeting policies. Would the transition to strategic defense make both sides more secure, or would each side alter its nuclear employment policies in such a way that cities and population centers would face even greater danger than they have in the recent past?

The Irony of Strategic Defense

If arms control agreements succeed in reducing the levels of offensive nuclear weapons, there still will remain a visible trend toward modernization and qualitative advances in the remaining forces. Maneuverable warheads, Stealth technology, and cruise missiles, to name a few, will be sufficient to create doubts about the effectiveness of defenses. Similarly, technological breakthroughs in defenses will increase the uncertainties for offensive operations. Together, offensive and defensive uncertainties may precipitate targeting policies that are as threatening as any in the past. Cities and their civilian populations could again become primary targets in a nuclear war. This outcome would be the ultimate irony of strategic defense.

Table 1 illustrates the relationship between current U.S. strategic doctrine based on limited attack options and the evolution toward a defense-dominant world. The phases are based on Nitze's descriptions. During the initial decade, assuming a cooperative adversary, offensive

Table - 1

THE NEW STRATEGIC CONCEPT

Impact on U.S. Strategic Doctrine

1985————————1995 Reduce Offensive Forces	1990————————1995 Interim Point Defense	1995————————2015 Territorial Defense	2015————————? "Ultimate Period" Near Zero Nuclear Offensive Forces
Limited Nuclear Options Remain Credible	Reduced Credibility Against: • Strategic nuclear targets • Leadership targets • Some conventional targets Most Credible Against: • Urban/Industrial • Transportation • Energy • Population	Limited Nuclear Options Not Credible	Offensive Nuclear Doctrine not Required
		Offensive Remedies: • Rely on Deterrence Through Defense • Technological Modernization to Penetrate Defense • Increase Offensive Forces • Attack High Value, Soft Targets If Deterrence Fails	

nuclear forces would be reduced (and modernized) to mutually agreed levels. Limited attack options could remain credible throughout this period.

If the United States were to begin deploying interim point defenses in the 1990s, such defenses would be deployed to defend strategic forces and command and control centers. Assuming that the Soviets deployed point defenses with the same priorities, a strategy of limited attack options would have significantly reduced credibility against strategic nuclear and leadership targets. Urban-industrial targets would become the most vulnerable target sets during a transition stage with extensive point defenses.

As point defenses then expanded to full-scale, space-based defenses capable of providing reasonably credible (but less than perfect) territorial defense, the credibility of limited attack options would be degraded against all target categories. As Table 1 indicates, however, several offensive countermeasures are possible. Ironically, urban-industrial targets may become the most vulnerable to attack in a less-than-perfect territorial defense. Offensive planning, unless all warfighting strategies were foregone, would avoid attacks against high-expenditure/low-payoff military targets, especially those that are hardened and protected by "thick" terminal defenses. Admittedly, a nuclear attack would come only in the most desperate crises; but if such an attack occurred, targeting plans would likely call for attacks on soft targets where a small, surviving force would have high payoff in its destructive effects. Only urban-industrial and nondispersed conventional military targets meet these criteria. The probability of nuclear war may decline in direct proportion to quantitative and qualitative offensive constraints, but the possibility of war will never reach zero. If deterrence does fail, the consequences might well be catastrophic, given the assumptions that each side might make about the other's defenses. The irony of strategic defense is that cities and population centers very likely could move from the bottom to the top of targeting priorities for both the United States and the Soviet Union.

Proponents of strategic defense argue that it is better to protect the U.S. people from nuclear attack than to avenge them in the event of an attack. There is a danger, however, as the preceding arguments suggest, that oversimplified appeals may detract from important issues of military strategy in a defense-dominant world. Deterrence can never rest on defense alone. Without offensive teeth, failure becomes the only penalty for aggression. The threat of offensive retaliation in some form will and should remain part of U.S. strategic doctrine. The strategic defense debate has raised serious questions about what form the offensive component of U.S. strategic doctrine may take in the twenty-first century.

New technology is not likely to be a substitute for offensive strategy or even for the classical theories of war. Clausewitz, for example, in his study of war, which has become a standard text in the curriculum of U.S. war colleges, devoted extensive effort to the analysis of defense. He concluded that defense was a stronger form of war than was attack. But defense was not purely passive. In his view, defense consisted of two phases: waiting for a blow and parrying it. The latter and sometimes forgotten action was intrinsic to Clausewitz's whole concept of defense. An army took up defensive positions in order to fight from them. A defense was a shield, but an active shield, one "made up of well-directed blows."[10]

Clausewitz's defensive strategy consisted of finding a proper balance between defense and offense, waiting and countering appropriately, and choosing the right time and place to unleash that "flashing sword of vengeance," which he described as "the greatest moment for the defender."[11] This view of the offensive-defensive relationship represents classic military thinking. No one wins war with defense alone. Wars are won or deterred by an army's ability to conduct offensive operations that destroy or threaten to destroy the military force of its enemy.

The SDI debate has reflected very little of this kind of thinking. Rhetorical excesses have created the impression that new technologies may become so reliable that the United States will be able to sheathe its strategic sword and rely on its shield. Beating swords into satellites will not free the United States from the threat of nuclear war. Offensive forces will remain in one form or another. Before we jump enthusiastically on the SDI bandwagon, it is important that we examine these realities and the offensive-defensive relationship during the projected transition period to a defense-dominant world. We need to know where that bandwagon is headed and what the offensive-defensive relationship may look like down the road.

The Impact of Strategic Defense on Soviet and U.S. Strategic Doctrines

Classically, deterrence of war and strategic nuclear weapons employment policies have a paradoxical relationship—deterring nuclear war has required policies, credible plans, and strategies that at least assure that potential adversaries could not win. Although winning a nuclear war has little meaning in view of the major destruction that would accompany the use of nuclear weapons, it still seems clear that to be deterred, potential aggressors must be denied confidence that they could achieve their war aims.

The strategic doctrines of the United States and the Soviet Union have evolved from this paradox, but there are important differences in emphasis.[12] U.S. force structure and declared employment policies have evolved to deter, through assured retaliation by survivable U.S. nuclear forces, Soviet execution of large-scale war plans or aggression in Western territory. Increasingly, this strategy has included damage limitation through preferential attack options against Soviet military targets that are accompanied by the threat of escalation to urban-industrial targets if aggression continues. (Damage limitations has two distinctly different meanings. One use of the term refers to selective attacks that limit collateral damage to the enemy. A second, more common use refers to preemptive attacks—i.e., attacks against enemy forces before they can be used.) From the U.S. perspective, the most salient feature of the damage limitation doctrine has been the evolution of graduated, flexible responses that incorporate limited nuclear attacks in order to maintain options for intrawar bargaining, escalation control, and prompt conflict termination.

Soviet doctrine places greater emphasis on warfighting and damage limitation through large-scale, preemptive attacks against military targets. Soviet force structure and declaratory policy emphasize that the society's capacity to survive a nuclear war depends on how well Soviet armed forces are prepared to fight a nuclear war. Moreover, the more clearly an adversary understands this preparedness, the more that adversary will be effectively deterred. This doctrine is sometimes called "deterrence through denial"—that is, seeking to deny the opponent the prospect of military victory. Deterrence through denial covers all Soviet strategic bases because it rests on well-established warfighting doctrines and capabilities in the event deterrence fails. U.S. strategists who favor warfighting options against Soviet military targets also argue that denial of military victory is a far more credible strategy than threats to punish an attacker by retaliating against civilian populations.

Although Soviet and U.S. strategic doctrines converge in their emphasis on hard-target counterforce and damage-limiting capabilities, potentially destabilizing doctrinal differences remain. The most obvious is the apparent Soviet rejection of limited nuclear war concepts, including escalation control and intrawar bargaining. The Soviets view these concepts as attempts at political intimidation rather than as elements in a strategy conceived by those who take war seriously. For the Soviets, denial of military victory requires robust preemption when war appears imminent and attacks of greater magnitude than those prescribed by U.S. limited nuclear war strategy.

Both doctrines are sensitive to the evolving relationships between offensive and defensive forces, and both are complicated by the fact that nuclear strategy is never purely defensive; the strategy in some respects invariably supports or lends credibility to offensive operations. Any calculation of a first strike or preemption is conditioned in part by active (e.g., ballistic missile defense, or BMD) and passive (e.g., civil defense) defensive capabilities to absorb residual second-strike forces. Any realignment of offensive and defensive strategic capabilities as envisioned by President Ronald Reagan's new concept of strategic defense must be examined carefully for its impact on the comparatively different offensive doctrines of the Soviet Union and United States.

A strategy incorporating strategic defense may or may not add to stability or to the evolving limited nuclear war capabilities of U.S. forces. Those outcomes will depend not only on the success and reliability of developing technologies but also on the Soviet Union's willingness to negotiate offensive limitations rather than to embark on new strategic initiatives of its own.

Precisely which general combinations of negotiated offensive-defensive constraints would degrade Soviet capabilities most is debatable. On balance, offensive reductions would affect the Soviet's robust style of preemption/damage limitation more than they would the evolving U.S. strategy of limited attack options and escalation control. Defensive constraints affect both U.S. and Soviet strategic doctrine. When combined with offensive limits, however, they degrade the effectiveness of Soviet forces more than those of the United States because defensive constraints make the execution of limited nuclear options more credible than does a Soviet strategy based on massive preemption.

Defensive advantages by either side will enhance greatly the credibility of that side's stratgic doctrine. Neither side therefore is likely to accede to a posture of defensive inferiority. In the event of failing arms control remedies, the disadvantaged party will seek to reestablish its strategic position through offensive countermeasures, defensive countermeasures, or both. In the Soviets case, these measures also could include doctrinal modifications. For example, the Soviets could seek compensation for perceived offensive shortfalls by moving toward a "softer" strategic target set, including greater emphasis on countervalue targets to compensate for the rapidly declining penetrability of Soviet strategic forces.

Offense, Defense, and Arms Control

The transition to a defense-dominant world through offensive arms reductions, even if the Soviets agreed, would confront difficult problems in maintaining offensive-defensive equilibrium. Treaties that result in equal offensive force levels may satisfy domestic political requirements, but they do not necessarily support the operational effectiveness of nuclear forces if deterrence fails. This is not to suggest that warfighting plans and strategies should drive arms control policy. Nevertheless, strategic force levels codified by treaty will shape warfighting options for the future, and the U.S. ability to deter war will depend to a large degree on the relationship between offensive trade-offs and defensive systems that may or may not be constrained by arms control agreements. For example:

- Strategies designed to negotiate higher U.S. bomber limits to trade against Soviet ICBMs must take into account offensive threats to bomber bases and defensive (air defense) threats to bomber penetration.
- To remain viable, submarines must be able to survive both offensive threats to their home ports and antisubmarine warfare (ASW) at sea, while their missiles must be able to penetrate enemy missile defenses.
- Because space-based ballistic missile defenses can be attacked by antisatellite weapons (ASATs) and possibly antiballistic missiles (ABMs), space-based BMD systems must be able to defend themselves and therefore must have the ability to destroy ASATs. Limitations on ASATs may enhance the survivability of space-based defenses, but either side could circumvent treaty limitations by

labeling an ASAT weapon as a BMD system or component. Conversely, BMD constraints could be circumvented by labeling a BMD weapon as an ASAT system or component. Because of their dual capabilities, both or neither should be constrained by treaty, but not one or the other.

- Similarly, space-based defenses could attack other space-based defenses. War in space therefore could begin with preemptive attacks by "defensive" systems against defensive systems.
- New technologies that may emerge from an unconstrained SDI could further obscure offensive-defensive relationships. If, for example, space-based "defenses" acquired a dual capability to destroy offensive weapons in flight and surface-based targets (e.g., intercontinental ballistic missiles [ICBMs], ABMs, ships), then "defensive" systems would not only support the offense indirectly by limiting a retaliatory attack but also directly through preemptive attacks against many types of targets.

These examples illustrate why nuclear arms control negotiations and military strategy are closely linked. Treaties cannot embrace every possible threat or contingency, but neither should they result in vulnerable force structures because arms control negotiators or military planners failed to comprehend the offensive-defensive relationships among strategic forces. There are no quick technological fixes to the dilemmas described here. However, the impact of strategic defense on arms control, offensive forces, and targeting policies that will remain in effect for at least the remainder of this century requires far more scrutiny than it has received in a debate that, thus far, has focused on public diplomacy, technical problems, and budgeting. There is a potential danger even in success if the SDI debate becomes too narrowly focused on technical breakthroughs. Technical successes may cloud judgments on nontechnical problems at a time when decisions are required to move from research and development to deployment. In evaluating the results of research and in taking any such decisions, the Reagan and subsequent administrations will need to ask some very basic questions about the future nature of U.S. strategy. In particular, they will have to consider how best to enhance deterrence and how best to curb rather than stimulate a new arms race. At that stage, the judgments to be made will depend only partly upon technical assessments about the feasibility of defenses. Even if the research shows promise, the case for proceeding will have to be weighed in the light of the wider strategic implications discussed in this chapter.

Notes

1. Department of Defense Press Release, December 19, 1984, p. 3.
2. Ronald Reagan, *The President's Strategic Defense Initiative* (Washington, D.C.: The White House, January 1985), p. 1.
3. Richard Nixon, *U.S. Foreign Policy for the 1970s: A New Strategy for Peace,* a report to Congress of 18 February 1970 (Washington, D.C.: Government Printing Office, 1970), p. 122.
4. James Schlesinger, Address for Writers Association Luncheon, DOD Public Affairs Office, 10 January 1974, pp. 5–6.
5. Desmond Ball, *Targeting for Strategic Deterrence* (London: International Institute for Strategic Studies, 1983), pp. 23–24.
6. Ibid., p. 24.
7. Ibid., p. 23.
8. Ibid. The radical shift in policy can be seen by comparing Reagan's strategic modernization speech of 2 October 1981 with his Star Wars speech.
9. Ambassador Paul H. Nitze, Statement before the Senate Foreign Relations Committee, 26 February 1985, pp. 2–5.
10. Carl von Clausewitz, *On War,* edited and translated by Michael Howard and Peter Paret (Princeton, N.J.: Princeton University Press, 1976), p. 357.
11. Discussed extensively by Michael Howard, *Clausewitz* (Oxford: Oxford University Press, 1983), pp. 23–26.
12. Studies of Soviet and U.S. strategic doctrine are too numerous to cite. Those that compare and contrast the Soviet and U.S. approaches to strategic doctrine include Fritz W. Ermarth, "Contrasts in American and Soviet Strategic Thought," *International Security* 3, no. 2 (Fall 1978): 138–155; Benjamin Lambeth, *Selective Nuclear Options in American and Soviet Strategic Policy,* R-20034-DDRE (Santa Monica, Calif.: Rand Corporation, 1976); and Dennis Ross, "Rethinking Soviet Strategic Policy: Inputs and Implications," *Journal of Strategic Studies* 1, no. 1 (May 1978): 3–30. For the evolution of U.S. strategic doctrine, see Fred Kaplan, *The Wizards of Armageddon: Strategists of the Nuclear Age* (New York: Simon and Schuster, 1983); and Leon Sloss and Marc Dean Millot, "U.S. Nuclear Strategy in Evolution," *Strategic Review* 12, no. 1 (Winter 1984): 19–28.

2

Is Ballistic Missile Defense a Good Idea?

Donald M. Snow

There is a piece of folk wisdom that contends, "If it ain't broke, don't fix it." Quoting such a homily may seem a strange and anomalous way to begin an inquiry into a subject as exotic and sophisticated as ballistic missile defense and President Ronald Reagan's Strategic Defense Initiative (SDI), but the homily applies nonetheless.

A small mountain of studies focusing mostly on the technologies that might make up the SDI has emerged since the president's major address of March 23, 1983.[1] The effect of that speech was to give great public visibility to research efforts begun at least a decade earlier and inherited by the Reagan administration from President Jimmy Carter (who had created the Directed Energy Transfer Office within the Department of Defense).[2] The president made front page news out of missile defense for the first time since the Antiballistic Missile (ABM) Treaty was signed in 1972.

The president's enthusiasm and progress in some aspects of the SDI have kept it at the top of the strategic agenda. The so-called Hoffman Report,[3] for instance, concluded that active defenses can make future contributions, and the administration has included defenses as a centerpiece of both its strategic and arms control agenda, as clearly demonstrated at the Geneva summit. Called the New Strategic Concept by chief arms control negotiator Paul H. Nitze,[4] the plan calls for a phased movement away from nuclear offensive weapons dominance to a world dominated by nonnuclear defenses against ballistic missiles. The New Strategic Concept is a three-stage process whose full implementation will extend to some undefined time in the twenty-first century; the concept is based upon the development of survivable defenses (invulnerable to attack) that are also cost effective (cheaper than offensive countermeasures). Such defenses "could so complicate a potential attacker's planning for a first strike that such an attack could not be seriously contemplated."[5] The plan resembles proposals such as Barkenbus and Weinberg's defense-protected build-down (DPB) of offensive weaponry.[6]

The resulting public and scholarly debate has focused on the SDI, and this has had both insulating and distorting effects. Focusing attention on the SDI insulates the debate because it is, in reality, a discussion and analysis about something that does not exist, may or may not ever exist, and, if it does, may exist in ways and with capabilities that cannot be fully predicted. It is, in other words, impossible to pass final judgment on a system whose characteristics are unknown, although an unfortunate amount of analysis has attempted to do precisely that. At this point, the SDI remains ethereal.

The SDI focus also has distorted the broader strategic debate and raises valid images of Buck Rogers warfare, the militarization (more properly, the weaponization because reconnaissance satellites already perform military purposes) of space, and impassioned rhetoric about whether laser or particle beams can ever perform, at what levels they can be made to work, and what arms race spiraling actions such systems will stimulate. These kinds of questions, common in the commentary on the SDI, unfortunately divert and distort what should be the more basic considerations.

Whether a laser beam can be made to destroy or disable a warhead or a strategic nuclear delivery vehicle (SNDV) is not the important question. The technologies that make up the SDI research program are, after all, nothing more nor less than instrumentalities designed to serve politically determined strategic objectives. What does matter is whether any form of ballistic missile defense (BMD)—as derived from the SDI initiatives in directed-energy transfer, antiballistic missilery, or any of the other "off-the-shelf" technologies that constitute the High Frontier[7]—enhances or contravenes U.S. strategic objectives. Those objectives have been defined for the past quarter century as deterrence of nuclear war. Thus, the fundamental issue is whether missile defenses reinforce or detract from deterrence. To return to the original analogy, is the deterrence system in need of repair (is it "broke")? If it is, will missile defenses fix what is wrong with it? If it works, why fix it?

The current dialogue, which gravely invokes the relative merits of x-ray lasers, rail guns, and various "architectures" (design plans) for a future system, almost totally ignores this most basic concern. The contribution of active defenses to deterrence certainly has been raised before but in a different technological context and with mixed, incomplete results. Glenn Snyder first muddied the issue in a landmark work in 1961 that defined deterrence and defense as antithetical, either/or concepts. According to Snyder, the former deals with war prevention and the latter with what happens after deterrence fails,[8] defenses are not seen in this context as a potential instrument of deterrence. The issue of defenses was further tainted in the early 1960s by its association with opposition to the Limited Test Ban Treaty because one of the few groups arguing for continued atmospheric testing was the scientific community investigating the properties of nuclear explosives against a missile attack.[9] The theoretical desirability of BMD was addressed in the ABM debate of the latter 1960s, but that discussion was largely overwhelmed by the acrimonious debate about whether ABMs, and particular ABM systems, were technically workable (an emphasis not dissimilar to the distortion in the current debate about whether the SDI is technologically feasible).

Raising the question of the need and desirability of defenses logically precedes consideration of the attractions of particular BMD schemes. If, as Snyder argued a quarter century ago, defense and deterrence are incompatible, then either the effort should be abandoned or some alternate basis for strategic weapons strategy must be devised that incorporates defense as part of deterrence. If some forms of defense reinforce deterrence and some do not, then those that do should be spotlighted and those that do not should be abandoned. If deterrence and missile defenses are comprehensively compatible, then a greater emphasis on BMD across the board seems warranted. What is clear is that some alteration in tradition is needed because deterrence thought has been dominated by the offense. As Douglass and Cohen put it, "The strategy needs to be redesigned. The reason is obvious. The current strategy does not allow for defense."[10]

The first step in the assessment process is to determine the state of the current debate. Two initial points stand out. First, current positions on the SDI debate have largely been framed in terms of preferences for traditional declaratory strategies that may inadequately encompass evolving reality. Second, the terms of the debate are changing rapidly in reaction to altered perceptions about technological prospects. Following some assessment of the current state of the debate about the desirability of defense, I will discuss the impact of missile defenses on strategic uncertainty, a factor increasingly recognized as central to the dynamics of U.S.-Soviet strategic interaction and then move on to the prospects of incorporating defenses with offensive force reductions through some form of arms control.

The Orthodox and Heterodox Positions

Since the beginning of the nuclear age, the U.S. strategic debate has been dominated by two major positions, both based at least implicitly on ideas published in 1946. One position emphasizes the horror and unacceptability of nuclear war in any form, the dictate for deterrence strategy based in the threat of punishment through awful retaliation, and the general view that nuclear war once begun would be uncontrollable and cataclysmic. This position was originally associated with Brodie[11] and found its most complete articulation in the strategy of assured destruction (AD) during the 1960s. AD as a policy has come under attack for a variety of reasons, and some of its advocates have moved from advocacy of AD as a preference to the subtler position that it is the unavoidable consequence of nuclear exchange. Robert Jervis captures one face of this distinction: "MAD [mutual assured destruction] as a fact is more important than MAD as a policy. The latter is in the realm of choice; the former is not."[12] Another more subtle variant is stated by Schelling, whose "hostage effect" concept is a key to AD thinking. "Since 1964 the correct name of the strategy is not 'assured mutual destruction' but 'assured *capability* for mutual destruction'" [emphasis in original].[13]

AD adherents historically have disliked missile defenses, although their hostility is greater to some defenses than to others. In essence, the AD philosophy accepts the defense of retaliatory forces as at least marginally useful ("deterrence enhancing" in one recent description) but views attempts to protect population as destabilizing ("deterrence escaping" according to the same author).[14] This distinction between protected objects has both theoretical and practical bases.

The theoretical base of this distinction is found in the hostage effect, which posits that deterrence rests on the knowledge that any nuclear aggression will lead to a devastating retaliation, thereby making the initial attack suicidal. Each population is a hostage of the other and will be executed for nuclear transgression.[15] Given this presumption, defenses of retaliatory force can increase retaliatory forces available after an aggression, thus impeding the ability to calculate (or miscalculate) survival. Population defense, on the other hand, may weaken inhibitory mechanisms derived from the hostage effect by allowing the (right or wrong) calculation that one might not be committing suicide by launching an attack.

Adherents of AD also tend to consider population defense an unattainable chimera, especially if city defense must be perfect to be desirable. Two former secretaries of defense, who gained experience in the McNamara Pentagon that spawned AD, have joined the chorus deriding the SDI as incapable of comprehensive population protection. Harold Brown, for instance, states, "American political and military leaders should publicly acknowledge that there is no realistic prospect for a successful population defence, certainly for many decades, and probably never."[16] More tersely, James Schlesinger asserts, "There is no realistic hope that we shall ever again be able to protect American cities. There is no leak-proof defense."[17]

As doubts about a leakproof population defense have increased, another aspect of the debate has surfaced—that is, "whether and to what extent the United States should deploy active defenses with more limited objectives and capabilities."[18] The AD position is ambivalent on this point. A partially effective system of retaliatory force defense is more valuable than none at all because some forces that would otherwise be destroyed will survive. However, many AD supporters fear that a partial system would be a kind of Trojan horse that, once begun, could inexorably expand to the opposed population defense. Also, some feel the costs associated with such a system outweigh the gains: "Deployment of limited or intermediary ballistic missile defenses would, on balance, reduce American security. . . . Deployment of less-than-perfect BMD and the associated evolutionary strategy should be rejected."[19] Costs cited include damage to the ABM Treaty, a race in offensive countermeasures, and increased difficulty in negotiating arms control agreements.

The other basic deterrence position also embraces deterrence as the major goal of nuclear weapons but disagrees on how to maintain it. Originally stated in a relatively unknown book in 1946 by Borden, this position argues that nuclear weapons make attacks on civilian targets unconscionable and unthinkable because of the consequences, thereby making all-out nuclear attacks incredible. Thus, more limited, measured hostilities are possible and require deterring. Borden himself states, "Victory in another conflict will not depend on destruction of civilians but on quick elimination of the opponent's forces and stock piles in being."[20] This statement of the problem emphasizes counterforce targeting and the ability to wage nuclear war at various levels of intensity and assumes adversary knowledge of that will and ability will be dissuasive. This policy preference has variously been called flexible or controlled response, limited nuclear options (LNOs), and countervailing strategy. Bearing the resuscitated title of flexible response, it is the deterrence strategy of the Reagan administration.[21]

BMD and the philosophy of LNOs are compatible, whether one is considering counterforce or countervalue targets. Speaking of protecting ICBMs and command, control, and communications assets, ex-national security adviser Zbigniew Brzezinski argues, "A limited strategic defense . . . would enhance deterrence by making it more difficult to calculate the advantage of a first strike."[22] In recognizing that deterrence can fail and that responsible persons must try to mitigate the disaster, Brzezinski contends defense is desirable because "a Soviet assured-destruction capability continues to exist as a result of the enduring official U.S. disinterest in strategic defenses, with potentially paralyzing implications for the United States."[23]

Supporters of the LNOs are quite sensitive to what they view as nuances in Soviet strategy, especially at the lower range of hostilities, that could provide the Soviets with advantages and that must be counteracted accordingly. Soviet professed interest in active and passive defense and the Soviets' long-standing disdain for AD fuel such a concern: "There is a more fundamental reason why we should not expect the Soviet leadership to settle for the strategic order of consensual vulnerability. . . . The accord on a stable equilibrium of mutual restraint is psychologically incompatible with the constant threat of reciprocal annihilation."[24] Moreover, there is concern that active defenses logically follow from the Soviet intention to win and survive a nuclear war should one be forced upon them. "The Soviets in principle have never denied the military utility of BMD in the context of nuclear war-fighting scenarios."[25] If the consistent Soviet opposition to the SDI seems to contradict an interest in defense, there is an answer to that, too: "The Soviet Union is not unfriendly to the idea of homeland defense, only to the idea of American homeland defense."[26]

The U.S. fear is that an uncountered, and therefore advantageous, Soviet defensive effort could compromise the United States in some future crisis and render U.S. deterrence threats less credible. This reasoning applies equally to protecting counterforce or countervalue targets, to research efforts as well as deployments, and to imperfect as well as perfect defenses. Any U.S. defensive capability greatly complicates Soviet attack scenarios; the Soviets would be unable to determine which potential targets could be protected. Even a less-than-perfect defense would preclude the demise of certain targets, thereby complicating the Soviet ability to calculate gain because "a condition of stable deterrence is one wherein Soviet leaders anticipate the defeat of their strategy."[27] As Kampelman states the goal, "I believe we would all agree that it would be better to base deterrence on an increased ability to deny the aggressor his objectives than to rely solely on our ability to punish him. . . . It is this prospect for a more effective deterrence that research on strategic defense offers."[28]

There is a third way to view the dynamics of deterrence—realistic self-deterrence (RSD). This view gives rise to different implications about BMD attractiveness.[29] The basic tenet of RSD is that the first foreign policy priority of both superpowers is the avoidance of nuclear war with the other.[30] This priority is the result of the mutual belief that such a war's consequences are intolerable; each side thus deters itself, and the secret of deterrence strategy is how to maintain that condition. This effort becomes at least partially cooperative—"because of this combination of genuine rivalry and genuine common interest, both superpowers are cautious even when pursuing policies designed to minimize the likelihood of war between them."[31] As a result, the principal problems are crisis avoidance management and the avoidance of circumstances that would generate perceptual changes regarding the tolerability of war.

The impact of BMD can be assessed against these dual criteria. Crisis management/avoidance seeks to ensure that circumstances do not arise that inadvertently could lead to war, or, if they do, that they are quickly and harmlessly defused. Defenses of population would not affect this goal greatly, but retaliatory force protection could be beneficial: "Development of strategic defenses . . . is important for the U.S. . . . because it cancels an advantage—the current superiority of Moscow in heavy land-based missiles—that could encourage the Kremlin to take risks leading to conflict with the U.S."[32] This is, however, an area that clearly calls for further investigation because, "crisis prevention may well be considered the orphan of strategic studies."[33]

At first blush, the impact on perceptual changes would appear to depend on whether deployments were symmetrical or one-sided and whether the systems were less than perfect or not. The greatest possibility for recalculating the tolerability of nuclear exchange would occur if one side could field a perfect system when the other had no system, and the smallest possibility would accompany mutual deployment of imperfect systems. The other possibilities (both field perfect systems; only one can field a system that is imperfect) fall somewhere in between in terms of effect. Once again, these dynamics merit more complete investigation.[34]

I'm Not Sure: Uncertainty on BMD

Recognition that great uncertainties exist about what would happen in a nuclear war has become increasingly evident in the recent literature. The central theme of the literature is that great operational uncertainties surround the conduct of nuclear war, something akin to a Clausewitzian "fog and friction," and these uncertainties cloud efforts to calculate gain from potential nuclear hostilities, thereby enhancing deterrence. I, not without criticism, have argued that strategy incorporate uncertainty enhancement as a major emphasis; such uncertainty certainly aids in avoiding perceptual changes about the utility of nuclear war within realistic self-deterrence.[35]

Strategic uncertainty as policy has not been embraced unanimously. Dyson, for instance, argues that the idea runs contrary to U.S. nuclear thought, which "demands certainty while Soviet strategy accepts uncertainty as inherent in the nature of war."[36] At the same time, Altfeld maintains that uncertainty may inhibit a cautious decisionmaker in calm circumstances but might be quite ineffective in deterring a high risktaker during a crisis.[37] "The Strategic Defense Initiative . . . raises many uncertainties and risks for U.S. deterrence strategy."[38]

Beneficial or not, the prospect of missile defenses, and especially those based in exotic technologies such as the SDI, creates "a new world of uncertainty and unforeseen consequences."[39] Those who maintain that missile defenses benefically enhance uncertainty point to the complications even a partially effective system creates for the Soviets. "The primary purpose of the U.S. strategic defense is to increase Soviet planners' uncertainties about the prospects for successfully executing Soviet strategy."[40] The reason for this increased uncertainty is that Soviet planners could never calculate which targets (especially retaliatory systems) would survive and what systems to target for follow-on attacks. The result of partially successful defenses is to make it impossible to determine how much of the attacked force would be available for retaliation and thus what the costs of attack are. This uncertainty, the argument goes, enhances stability and deterrence. "Terminal non-nuclear defenses could provide a very attractive and effective way of strengthening strategic stability. Such defense would have to promote massive new uncertainties in Soviet attack calculations—calculations that already are beset with major technical, tactical, strategic and political uncertainties."[41] This helps Gray conclude, on balance, that the SDI is desirable: "The SDI-technical uncertainties and novel strategic problems admitted offers the only path that may be available to lead toward our living in much greater safety."[42]

Not everyone agrees that adding BMD to the strategic mix would be beneficial. In addition to objections such as Altfeld's about whether uncertainty enhancement deters at all, there is the additional issue of whether any incremental benefit derives from BMD. Glaser, for instance, feels there is no additional benefit. "A decision-maker who could be deterred by uncertainties about attack outcomes would likely be deterred by the uncertainties that already exist. In contrast, if these uncertainties in combination with the other reasons for not launching a counterforce attack are not adequate to deter a Soviet decision-maker, then the additional uncertainty created by BMD is likely to have little effect.[43]

More Stock for the Cauldron: Offensive Reductions and Arms Control

A critical aspect of the relation of missile defenses to nuclear war avoidance is the effectiveness of the defense; variation in defense helps determine which potential targets defenses can protect. In its roughest form, partially effective defenses can be used to defend retaliatory systems because the requirements for such systems are incremental; to whatever degree these defenses are effective, more systems survive, and there are incremental survival benefits as they become better. At the other extreme, the ideal population defense must be perfectly leakproof because even a few offensive forces penetrating the defenses would cause massive, probably unacceptable death and destruction. Between the extremes, a perfect defense of retaliatory forces is incrementally better than a less-than-perfect one, but it is highly debatable whether an imperfect population defense is better than none at all.[44]

The problem, of course, is that less-than-perfect defenses are considerably easier and less costly to build than perfect systems, which may not be possible to build at all, as earlier skeptical assessments suggested. Early advocates of the SDI raised hopes that comprehensive, airtight defenses might be attainable, and there are still some supporters of this position.[45] Much of the skepticism about these prospects concerns defensive peformance against an unconstrained offensive force that can simply expand to overwhelm any possible defense (assuming the offense remains cheaper than the defense, in contravention to one of Nitze's New Strategic Concept criteria).

A growing number of observers are coming to agree either that the defense case is strengthened by constraining offenses or that truly effective defenses are impossible without reductions or constraints on growth in offensive forces. Speaking to the former concern, Lord maintains that "the principal constraint on further strengthening of Soviet offensive forces is likely to be supplied by a serious U.S. commitment to strategic defense rather than by arms control as such."[46] Schlesinger underscores the mutuality of relationship between offensive constraint, defensive effectiveness, and a more stable situation in his summary of the literature: "A number of studies have been done that, I believe, effectively demonstrated . . . that there are certain cases in which a mixture of defense and offense would improve the position of the United States, the position of the Soviet Union, improve world stability, and provide a strategic relationship in which . . . there would be less damage. *All of these studies rest on the assumption that the defense is constrained*" [emphasis in original].[47]

The need to combine arms control negotiated offensive constraints and defensive introductions currently places the United States and USSR at loggerheads. Mikhail Gorbachev has been adamant that progress in offensive reductions requires abandoning what the Soviets refer to as "space-strike" weapons, a position absolutely at odds at least with those aspects of the SDI that could be deployed in space. Despite this political difficulty between the superpowers, the need to tie defensive deployments to arms control–imposed restraints is shared increasingly by defenders of defensive systems. Holloway, who is generally not enthused by defenses, states, "If defensive systems are to contribute to a safer and more stable strategic relationship between the United States and the Soviet Union, they will have to be imbedded in a strict arms-control regime that limits offensive systems."[48] Kruzel amplifies this assessment, with direct reference to SDI: "Limiting defense forces should theoretically make offensive forces less attractive, but limiting offensive forces will make defensive forces more attractive. . . . The only way strategic defense will ever be a good investment is against a constrained offensive force. . . . SDI does not make sense without a START [Strategic Arms Reduction Talks] agreement."[49] Gray believes that the Soviets will eventually come around to a movement toward defense, and the transition will be "managed by arms control."[50]

If one makes the dual assumptions that missile defenses are desirable and that their advantage is based on some offense-constraining arms control agreement, then it is necessary to ask if the arms control process is robust enough to produce the desired result. In the hiatus between Senate failure to ratify SALT II and the resumption of talks in Geneva during 1985, there was great skepticism about the ongoing role of arms control processes to regulate or reshape any aspect of the nuclear arms competition, much less anything as momentous as a transition to a defense-dominated world.

What is the basis for this skepticism? Four elements stand out. First, arms control was generally oversold during the early and middle 1970s. Arms control became the centerpiece of U.S.-Soviet relations and the beacon guiding us all away from the abyss of Armageddon. Arms control was elevated beyond its reasonable role, a situation that precluded a more modest position where "arms control is unlikely to produce either miracles or calamities. Kept in that perspective, formal arms control has a modest but useful role to play in an overall strategy of avoiding nuclear war."[51] Arms controllers, in other words, set standards that were excessive and thus assured the failure of the process by definition.

Second, there was suspicion that many of the advisers surrounding the president (if not the president himself) actually opposed arms control. Slocombe, for instance, argues, "Paradoxically, the upswing in the general popularity of reaching arms control agreements with the Soviet Union comes in the midst of the ascendancy . . . of people who have been throughout their political careers strong critics not just of the details, but of the very concept of arms control agreements."[52] Krepon echoes this sentiment by stating flatly that "key executive positions are held by ideologies opposed to arms control agreements."[53] Schelling, speculating on the current climate of proarms control rhetoric coming from the Reagan Administration, suggests the possibility of disingenuousness: "Perhaps an Administration with no genuine interest in arms limitation finds in arms control the best pulpit from which to preach arms competition."[54]

Third, arms control was reified by the government; it was not presented to the public as a reflection of U.S.-Soviet relations and mutual views toward nuclear war. Arms control is not a thing in and of itself with its own life; rather, it is a process and mechanism that the superpowers sometimes employ. As Gray covertly points out, "Progress in arms control reflects the quality of political relationships."[55] When those relations allow for success, it is achieved. When relations are cool or sour, prospects dim. Thus, Schlesinger's recent statement that "in Geneva we must prepare for an extended period of siege warfare—with the Soviets well positioned to exploit differences between our allies and ourselves"[56] is really a commentary on superpower relations, not on arms control as such.

Fourth, negotiating an offensive build-down, defensive buildup is a complex activity. Even if the political will were present, the technical difficulties involved would be staggering. When one considers that it took about five years to negotiate what the Reagan administration has frequently alleged as the "fatally flawed" SALT II document on offensive arms alone, then when defensive forces, intermediate nuclear forces, and, in all likelihood, conventional force levels are added to the agenda the difficulties increase exponentially. It is a staggering prospect.

What do these problems suggest about the contribution of arms control processes in the transition to a world that includes strategic defenses? The most obvious conclusion is not to expect too much from such processes. They cannot create order from chaos or amity from confrontation; rather, they are at best fragile forums wherein the superpowers sometimes negotiate their complex and conflicting interests. When arms control "triumphs," it means those matters being negotiated were ultimately resolvable and those doing the negotiating were clever or patient enough (or both) to exploit available opportunities. When arms control "fails," the ability or the will—or both—to reach agreement are missing or inadequate.

The current Geneva round of arms talks is a study in contrasts. On the one hand, unparalleled opportunity abounds. A new Soviet leader has made apparently bold proposals for offensive force reduction not dissimilar in potential effect to the oftstated aspirations of the U.S. president. Through the segment devoted to space, the SDI defenses are part of the agenda as well. All the elements for offensive build-down/defensive buildup are on the table.

On the other hand, the real intent of either side is not entirely clear; nor is the basis for agreement clear. Initially, at least, Gorbachev insists offensive reductions must be tied to abandoning at least that part of defenses that are space-based options within the SDI, and one wonders what weapons he is really willing to dismantle (e.g., only obsolete systems) or whether the offer is no more than a propaganda ploy. At the same time, the Reagan administration's real commitment to arms control always has been the subject to some suspicion, and USSR intransigence about the SDI could be used to subvert whatever real prospects there might otherwise be.

Even if the will to negotiate were present, the ability may not be. The ability to reach agreement is hindered by the breadth (number of items) and depth (technical complexity) of the agenda, which is further complicated by the interconnecting relationships between agenda items and the increasingly apparent asymmetries of the forces on which surgery is proposed. Fitting all the pieces together may simply be beyond human diplomatic capacity. Brzezinski may well be correct in saying that "one should not expect again a comprehensive agreement on the mode of SALT I or on the mode of the even more ambitious SALT II agreement. More likely are segmented, narrow agreements focused on specific aspects of the problem."[57] Kupperman suggests such a more limited approach: "One path toward stable arms reductions would be to employ B-14s as a buffering mechanism against violations of arms control agreements."[58]

This does not imply that arms control processes cannot be used to contribute to movement toward a defense-inclusive world. Such a movement suggests the very real problem of an orderly, nonthreatening transition that does not destabilize the balance, the "how-do-we-get-from-here-to-there" problem.[59] It is certainly a potential SDI problem because "the Soviet Union may consider initial U.S. deployment of BMD systems or components in space to be so inimical to Soviet security that they may decide to disable them."[60] At the same time, Gray may be correct that looking at SDI in arms control terms at this point is premature and misguided: "It is both bizarre and not a little sad that the current debate . . . suggests that the most important question to be asked of space weaponry of different kinds is how best to control it."[61]

Conclusion

Attempting to judge the impact of ballistic missile defenses on nuclear deterrence is not an easy business. Much of the difficulty is that the subject is enshrouded in quasi-theological assessments about what deters, and there are deeply held disagreements about the effect on deterrence of protecting different kinds of targets with differing levels of efficiency. Moreover, there is considerable disagreement on how to make the transition from a defenseless world to one in which defenses play a part.

In order to come to grips with the basic problem, three questions must be answered. First, is ballistic missile defense possible, and if so, how effective can it be? This is the necessary first question because a negative answer implies that there is no positive or negative effect on deterrence.

The jury is out on this question. As indicated earlier, there is considerable expert consensus that a perfect defense is unlikely or impossible in the absence of offensive constraints, but that conclusion is not universally accepted. As Edward Teller obliquely suggests, "Because of secrecy, the best publicized systems are those that are rather primitive or will not work at all."[62] "Whenever a truly effective idea is pursued, that is kept secret."[63] One is left with an interim conclusion—that offense-constrained defense with high effectiveness may be possible and that less-than-perfect defenses are plausible. At the same time, these systems would not provide comprehensive solutions to the national security dilemma because "the President's program offers no promise of effective defenses against anything but ballistic missiles. . . . There would remain the difficulty of defense against cruise missiles, against bomber aircraft, and against the clandestine introduction of warheads."[64]

If one concludes that some form of defense is technically possible, the next question is will the United States seriously pursue those prospects to the point of deploying defensive systems. Put another way, is the challenge of potential missile defenses theoretical and academic, or is it real?

This is a political and economic problem that is an important qualification with regard to the SDI. In his March 23, 1983, speech the president drew an analogy between the SDI effort and the Apollo program, but Kiser notes that "the political process is not likely to give the same support to strategic weapons that it gave to the moon race. At least four administrations and eight congresses must give strong support to the SDI for it to realize its ambitious aims."[65] Pure political partisanship is likely to corrode support during that extended period of time, and costs are likely to be very high (as much as $1 trillion is often cited).[66] This level of spending could simply make the SDI economically, and thus politically, unattainable, especially if the system did not include currently absent air defenses against bombers and cruise missiles. Simes maintains that the Soviets, despite their harsh public denunciations of SDI, realize this problem: "The Soviets . . . are equally aware that the costs are going to be monumental and that there is no certainty that Congress would approve them."[67]

The final question is whether missile defenses, if practicable and affordable, are desirable in the context of deterrence. Is the system "broke," and does it need fixing? To look at that question rationally and dispassionately requires stripping way the heat and glitter of the Star Wars debate from the more basic concern about whether defenses in any form are desirable.

Such a separation helps elucidate several issues. The first involves reversing the order of consideration of desirability and cost. The extravagant cost of the SDI may make its desirability moot because SDI is unaffordable. The danger is in throwing the BMD baby out with the SDI bath water. Considering the generic problem before its manifestations allows one first to decide if BMD is desirable and, based on that assessment, how much one can afford. That aspect of the question has been inadequately studied.

A second issue is to clarify the Soviet place in the equation. The Soviets are vehement about space-based SDI, and that vehemence alone could lead to rejection of the defense on the basis that it takes two to create stability. Viewing the question generically modifies that concern because despite their dislike for SDI, "the Soviets . . . have never claimed that a situation in which each superpower's territory is largely invulnerable to strategic attack necessarily upsets deterrence."[68] The Soviets, in other words, support BMD but not SDI.

In its most comprehensive, effective, possible forms, a movement toward an SDI-based world of missile defenses would radically alter the world of nuclear weapons and require basic reconsideration of the way the deterrence system is conceived and implemented. Given that through "trial and error the superpowers have developed some prudent practices"[69] for dealing with one another, anything that might change their relationship in fundamental ways must be viewed carefully and cautiously. Schelling is quite correct when he says of the system that "we have what ought to be an important source of reassurance, a 'confidence-building' experience: 40 years of nuclear weapons without nuclear war."[70] The system is not "broke," but that does not mean it cannot be improved. The real question is whether missile defenses will improve the world, and that must be answered definitively before we proceed.

Notes

1. The author's primary "contribution" to the debate is Gary L. Guertner and Donald M. Snow, *The Last Frontier: An Analysis of the Strategic Defense Initiative* (Lexington, Mass.: Lexington Books, 1986).

2. The president's chief negotiator, Max M. Kampelman, made a rare administration admission of this inheritance recently in a speech published as "SDI and the Arms Control Process," *Atlantic Community Quarterly* 23, no. 3 (Fall 1985):225.

3. Fred S. Hoffman, *Ballistic Missile Defenses and U.S. National Security,* Future Security Strategy Study (Washington, D.C.: Government Printing Office October 1983). For an update on Hoffman's views, see his testimony before the Subcommittee on Strategic and Theater Nuclear Forces of the Senate Armed Services Committee on March 1, 1985, reprinted as "The SDI in U.S. Nuclear Strategy," *International Security* 10, no. 1 (Summer 1985):13–24.

4. The administration's initial position at Geneva has been published as Paul H. Nitze, "Arms Control: The First Round in Geneva," *Current Policy,* no. 698 (Washington, D.C.: U.S. Department of State Bureau of Public Affairs, May 1985). This position is summarized by Gary L. Guertner in "Offensive Doctrine in a Defense-dominant World," *Air University Review* 37, no. 1 (November/December 1985):8–9.

5. Paul H. Nitze, "SDI: Its Nature and Rationale," *Atlantic Community Quarterly* 23, no. 3 (Fall 1985):265.

6. Alvin M. Weinberg and Jack N. Barkenbus, "Stabilizing Star Wars," *Foreign Policy,* no. 54 (Spring 1984):164–170. For some assessment of this idea, see, for instance, Alton Frye, "Strategic Build-Down: A Context for Restraint," *Foreign Affairs* 62, no. 2 (Winter 1983/1984):293–317; or Donald M. Snow, "BMD, SDI and Future Policy: Issues and Prospects," *Air University Review* 36, no. 5 (July/August 1985):11–12.

7. For a layperson's view of the High Frontier, see Daniel O. Graham and Gregory A. Fossedal, *A Defense that Defends: Blocking Nuclear Attack* (Old Greenwich, Conn.: Devin-Adair Publishers, 1983).

8. Glenn N. Snyder, *Deterrence and Defense: Toward a Theory of National Security* (Princeton, N.J.: Princeton University Press, 1961).

9. Freeman Dyson, *Weapons and Hope* (New York: Harper and Row, 1984), p. 74.

10. Joseph D. Douglass, Jr. and Samuel T. Cohen, "SDI: The Hidden Opportunity," *Defense Science 2003+* 4, no. 4 (August/September 1985):6.

11. Bernard Brodie, ed., *The Absolute Weapon: Atomic Power and World Order* (New York: Harcourt Brace, 1946).

12. Robert Jervis, "MAD is the Best Possible Deterrence," *Bulletin of the Atomic Scientists* 41, no. 3 (March 1985):43.

13. Thomas C. Schelling, "What Went Wrong with Arms Control," *Foreign Affairs* 64, no. 2 (Winger 1985/1986):230.

14. Alton Frye, "Strategic Synthesis," *Foreign Policy,* no. 58 (Spring 1985):9.

15. Thomas C. Schelling, *The Strategy of Conflict* (Cambridge, Mass.: Harvard University Press, 1960).

16. Harold Brown, "The Strategic Defense Initiative: Defensive Systems and the Strategic Debate," *Survival* 27, no. 2 (March/April 1985):63.

17. James R. Schlesinger, "Rhetoric and Realities in the Star Wars Debate," *International Security* 10, no. 1 (Summer 1985):5.

18. Brown, "The Strategic Defense Initiative," p. 55.

19. Charles L. Glaser, "Do We Want the Missile Defenses We Can Build?" *International Security* 10, no. 1 (Summer 1985):27–28.

20. William Liscum Borden, *There Will Be No Time: The Revolution in Strategy* (New York: Macmillan, 1946), p. 218.

21. See, for instance, Caspar W. Weinberger, *Annual Report to the Congress, Fiscal Year 1986* (Washington, D.C.: Government Printing Office, February 4, 1986), pp. 45–48.

22. Zbigniew Brzezinski, "East-West Relations," *The Atlantic Community Quarterly* 22, no. 4 (Winter 1984/1985):297.

23. Colin S. Gray and Keith Payne, "Victory Is Possible," *Foreign Policy*, no. 39 (Summer 1980):19.

24. Fred Charles Ikle, "Nuclear Strategy: Can There Be a Happy Ending," *Foreign Affairs* 63, no. 4 (Spring 1985):817.

25. David B. Rivkin, "What Does Moscow Think?" *Foreign Policy*, no. 59 (Summer 1985):100.

26. Colin S. Gray, "Space Arms Control: A Skeptical View," *Air University Review* 37, no. 1 (November/December 1985):85.

27. Ibid., p. 82.

28. Kampelman, "SDI," p. 224.

29. For a more elaborate statement, see Donald M. Snow, "Realistic Self-Deterrence: An Alternative View of Nuclear Dynamics," *Naval War College Review* 39, no. 2 (March/April 1986):60–73.

30. For a sample of official statements of this priority, see the following: Ronald Reagan, "The U.S.-Soviet Relationship," *The Atlantic Community Quarterly* 22, no. 1 (Spring 1984):5; George Shultz, "U.S.-Soviet Relations in the Context of U.S. Foreign Policy," *The Atlantic Community Quarterly* 21, no. 3 (Fall 1983):202; Gancho Ganev, Yevgeny Kazar, and Sarada Mitra, "The Nuclear Threat and Politics," *World Marxist Review* 27, no. 4 (April 1984):48; Nikolay V. Ogarkov, "Military," in Harriett Fast Scott and William F. Scott, eds., *The Soviet Art of War: Doctrine, Strategy, and Tactics* (Boulder, Colo.: Westview Press, 1982), p. 248; and Konstantin Chernenko, quoted in *Whence the Threat to Peace*, 3rd ed. (Moscow: Military Printing House, 1984), p. 14.

31. Harvard Nuclear Study Group, *Living with Nuclear Weapons* (Cambridge, Mass.: Harvard University Press, 1983), pp. 17–18, 238.

32. Carnes Lord, "The U.S.-Soviet Strategic Nuclear Relationship," in W. Bruce Weinrod, ed., *Confronting Moscow: An Agenda for the Post-Detente Era* (Washington, D.C.: The Heritage Foundation, 1985), p. 6.

33. Alexander L. George, "Toward a Soviet-American Crisis Prevention Regime," in Bernard Brodie, Michael D. Intriligator, and Roman Kolkowitz, eds., *National Security and International Stability* (Cambridge, Mass.: Oelgeschlager, Guin and Hain, 1983), p. 189.

34. For a useful introduction, see Stephen J. Cimbala, "The Strategic Defense Initiative: Political Risks," *Air University Review* 37, no. 1 (November/December 1985):28–35.

35. Donald M. Snow, *The Nuclear Future: Toward a Strategy of Uncertainty* (Tuscaloosa: University of Alabama Press, 1983), especially ch. 2, 4.

36. Dyson, *Weapons and Hope,* p. 189.

37. Michael F. Altfeld, "Uncertainty as a Deterrence Strategy: A Critical Assessment," *Comparative Strategy* 5, no. 1 (1985):1–26.

38. Cimbala, "The Strategic Defense Initiative," p. 24.

39. John W. Kiser, "How the Arms Race Really Helps Moscow," *Foreign Policy,* no. 60 (Fall 1985):51.

40. Lord, "The U.S.-Soviet Strategic Nuclear Relationship," p. 6.

41. Colin S. Gray, "Strategic Defenses: A Case for Strategic Defense," *Survival* 27, no. 2 (March/April 1985):52.

42. Gray, "Space Arms Control," p. 78.

43. Glaser, "Do We Want the Missile Defenses We Can Build?" pp. 36–37.

44. See, for instance, Snow, "BMD, SDI and Future Policy."

45. For recent statements, see, for instance, George F. Chapline, "In Defense of SDI," *Defense Science 2003+* 4, no. 4 (August/September 1985):14; slightly less positively, Robert H. Kuppermann, "Using SDI to Reshape Soviet Strategic Behavior," *Washington Quarterly* 8, no. 3 (Summer 1985):77; and Barry R. Schneider, "U.S. Military Space Policy in 2010," *Defense Science 2003+* 4, no. 4 (August/September 1985):28.

46. Lord, "The U.S.-Soviet Strategic Nuclear Relationship," p. 9.

47. Schlesinger, "Rhetoric and Realities," p. 7.

48. David Holloway, "The Strategic Defense Initiative and the Soviet Union," *Daedalus* 114, no. 3 (Summer 1985):277.

49. Joseph Kruzel, "What's Wrong with the Traditional Approach?" *Washington Quarterly* 8, no. 2 (Spring 1985):130.

50. Gray, "Strategic Defenses," p. 53.

51. Graham T. Allison, Albert Carnesale, and Joseph S. Nye, "An Agenda for Action," in Graham T. Allison, Albert Carnesale, and Joseph S. Nye, eds., *Hawks, Doves and Owls: An Agenda for Avoiding Nuclear War* (New York: Norton, 1985), p. 242.

52. Walter Slocombe, "An Immediate Agenda for Arms Control," *Survival* 27, no. 5 (September/October 1985):206.

53. Michael Krepon, "Dormant Threat to the ABM Treaty," *Bulletin of the Atomic Scientists* 42, no. 1 (January 1986):31.

54. Schelling, "What Went Wrong," p. 229.

55. Gray, "Space Arms Control," p. 75.

56. James Schlesinger, "The Eagle and the Bear: Between an Unfree World and None," *Foreign Affairs* 63, no. 5 (Summer 1985):961.

57. Brzezinski, "East-West Relations," p. 297.

58. Kupperman, "Using SDI," p. 83.

59. Donald M. Snow, "BMD, SDI and Future Policy," pp. 9–10.

60. Keith B. Payne, "The Strategic Defense Initiative and ICBM Modernization," in Barry R. Schneider, Colin S. Gray, and Keith B. Payne, eds., *Missiles for the Nineties: ICBMs and Strategic Policy* (Boulder, Colo.: Westview Press, 1984), p. 65.

61. Gray, "Space Arms Control," p. 75.

62. Edward Teller, "Defense: Retaliation or Protection," in Richard F. Staar, ed., *Arms Control: Myth Versus Reality* (Stanford, Calif.: Hoover Institution Press, 1984), p. 106.

63. Edward Teller, "Science and Technology in SDI," *Defense Science 2003+* 4, no. 2 (April/May 1985):17.

64. McGeorge Bundy et al., "The President's Choice: Star Wars or Arms Control," *Foreign Affairs* 63, no. 2 (Fall 1984):268.

65. Kiser, "How the Arms Race Really Helps Moscow," p. 50.

66. Snow, "BMD, SDI and Future Policy," pp. 10–11.

67. Dmitri Simes, "Are the Soviets Interested in Arms Control?" *Washington Quarterly* 8, no. 2 (Spring 1985):155.

68. Rivkin, "What Does Moscow Think?" p. 89.

69. Harvard Nuclear Study Group, *Living with Nuclear Weapons*, p. 238.

70. Schelling, "What Went Wrong," p. 219.

3

Strategic Defense: A Third View

Leon Sloss
Seymour Weiss

Since 1983 President Ronald Reagan's Strategic Defense Initiative (SDI) has been widely debated. Despite the millions of words that have been written and spoken on the subject, we believe that many key issues have been distorted or misunderstood. The Reagan administration, its critics, and the media all have contributed to a situation that has confused and polarized the issues.

The SDI presents the nation with important choices—choices that will affect the nature of U.S. nuclear strategy, the research, development, and production programs to support that strategy, and the relationship of both to arms control, for decades to come. The president's speech of March 23, 1983, was significant because it focused attention on defenses, a subject largely neglected in U.S. strategy for more than a decade. However, since that time the administration has failed to clarify the relationship between the SDI's long-range goal of abolishing nuclear weapons and less ambitious intermediate goals that have been alluded to but have not been clearly defined. Nor has the relationship between the SDI R&D program and the U.S. stance in the Geneva arms control negotiations been clearly spelled out. There is a potential conflict between SDI and the current Antiballistic Missile (ABM) Treaty, and this also needs to be faced directly.

Although technical issues have been debated extensively, the underlying strategic and policy issues that underlie SDI have not received the attention they deserve. The discussion that has occurred tends to posit a choice between two options—that is, between a traditional view of deterrence based on mutual assured destruction (MAD) in which defenses are bound to be destabilizing and a utopia in which all nuclear weapons are abolished and defenses are dominant. However, there is a third view. It is our intention in this chapter to argue that a balanced strategic posture with both offense and defense and an eventual revision of the ABM Treaty are required.[1]

No proposal with the technological and political complexities of the Reagan administration's Strategic Defense Initiative is likely to enlist unanimous support, or is such a proposal explainable in simple generalities. Although SDI should not receive support before a thorough public examination of its advantages and disadvantages takes place, we find it troubling that simplistic arguments, which confuse rather then enlighten, are advanced too frequently.

Both the critics and supporters of the SDI often distort the choices in order to make debating points. Many critics are more concerned with preserving the ABM Treaty than with exploring the possibilities of defense; some supporters of SDI evince little concern for arms control. The value of the ABM Treaty and the potential impact that the SDI may eventually have upon that treaty are legitimate issues for analysis. But a priori judgments concerning the sanctity of the former and the failure of the latter obscure, rather than advance, public debate. Surely there are circumstances in which continued adherence to the ABM Treaty in its present form would not serve U.S. national security interests. Conversely, there are circumstances in which a deployment of strategic defenses might make sense. If so, what are these circumstances?

We are told by the Gang of Four (Messrs. McNamara, Bundy, Smith and Kennan) that the SDI and arms control are incompatible.[2] Another gang, this time of six (Professors Bethe, Garwin, Gottfried, Kendell, Sagan, and Weisskopff), contend that an impermeable defense is impossible, and anything short of that "tends to undermine, not improve, U.S. security."[3] Sometimes even gangs of one, such as Professor Stanley Hoffman of Harvard, denounce the SDI because it will induce the Soviets "to concentrate on keeping our nuclear forces vulnerable," result in an inevitable multiplication of offensive capabilities, and, at the same time, force the Soviets to follow "America's example" by building a defensive system of their own.[4]

It is unclear on what evidentiary base these assertions rest. The Soviets have launched an extensive program of defenses. This would suggest they do not see a blatant inconsistency between defenses and arms control. Nor do they perceive defenses as inconsistent with Soviet security, and up to this point they do not appear to believe their defenses will lead to the inevitable multiplication of U.S. offensive systems. Perhaps the Soviets are misguided on all points, but surely the issue of U.S. security is important enough to warrant examination of the apparent premises underlying Soviet programs and how those programs ought to affect U.S. thinking. In any event, it would be surprising if the issues raised by critics of the SDI were in fact so simple, or so black and white.

Although it will become apparent that on many issues we diverge from the critics of the SDI, on one point we tend to agree with them. At this stage in our knowledge of advanced defensive technologies, it seems unlikely that we can create a perfect defense or completely eliminate nuclear weapons at any time in the foreseeable future. Even if the SDI were successful, it would not provide defense against all nuclear weapons. At some point it probably will be necessary to supplement the SDI with other appropriate defenses (e.g., air defense), and even if the elimination of all nuclear weapons from the face of the earth proved impossible their diminished use is still a worthy goal. In any event, although we share with critics their skepticism that the SDI can bring about a perfect, nonnuclear world, it does seem presumptuous for any of us to speak with great confidence about the future course of either technology or diplomacy several decades hence.

U.S. and Soviet Views of Strategy

One would think from a casual exposure to the current debate that defense was an entirely new idea and that a major revolution in strategy was imminent. In fact, the utility of defenses as a part of a sound strategy is not a new notion, nor is it correct to conclude the United States has a monopoly either on its theory or practice. Indeed, the utility of defenses is a matter of primary importance to the USSR, if one is to judge by Soviet actions rather than by contemporary Soviet public pronouncements. The debate that is now raging in the United States about the SDI is a replay, in a slightly new context, of arguments that have raged for decades.

Soviet Defense Initiatives

Many U.S. commentators, prominent among them the various "gangs" who are unreservedly critical of the SDI, believe that defenses were essentially abolished by the ABM Treaty of 1972 and that each side then laid itself bare to the devastating effects of an offensive strike. This view was never unanimously held, and it was never the view of the Soviet Union. Indeed, subsequent Soviet actions suggested an entirely different interpretation of the ABM Treaty—that it legitimized all forms of defense including antiballistic missile defense. Certainly, the Soviets have continued to pour massive resources into strategic defense, including an extensive research and development program for ballistic missile defense (BMD) that covers most of the technologies incorporated in the

U.S. SDI. They have deployed a limited ballistic missile defense, about which we will have more to say. In addition, they have maintained and modernized their air defenses. The Soviets have deployed 13,000 ground-to-air antiaircraft missile launchers in 1,400 fixed sites, and they have deployed 1,250 interceptor aircraft to protect the USSR, thereby far surpassing comparable U.S. and NATO efforts. They have a major civil defense program with 40,000 full-time personnel and a lesser-known but more substantial program to shelter large numbers of their political and military elites. They have built the hardest missile silos in the world and have developed a tactical ballistic missile defense system, which by definition has antistrategic ballistic missile capability. The Soviets possess the world's only operational antisatellite defense.[5] Secrecy, redundancy, dispersal, proliferation, camouflage, deception, and concealment on a scale inconceivable in the West—in short an enormous program of passive strategic defense—work to protect the entire Soviet political/military establishment.

The current Soviet ballistic missile defense is worth a further comment. While the U.S. dismantled the single ABM site permitted it by the treaty, the Soviets have—as allowed by the treaty—completed a defense around Moscow that, while nominally defending national political and military leadership, in reality provides some defense to a large portion of the industrialized USSR and numerous important military targets around Moscow.

It also appears to be the case that the vast Soviet surface-to-air missile (SAM) network may have some ABM capability as well, in particular the SA-10, a system the Soviets have tested to altitudes relevant to ballistic missile intercept. Another system, the SA-X-12, also is believed to have the capability to intercept some types of U.S. strategic ballistic missiles. We therefore are faced with some uncertainty as to just how modest the existing Soviet ABM capability is. Additionally, the Soviets have potentially created a production base for an expanded ABM system and have deployed components of a nationwide ballistic missile defense system (including massive radars that are a long-lead item in such a network). Finally, they have gained useful operating experience from existing deployments. Certainly all of this activity suggests that the Soviets are not uninterested in defense.

The ability of the Soviets to expand their current, first-generation missile defenses far more rapidly than could the United States (a capability they have achieved while the ABM Treaty has been in effect) serves as a deterrent to similar U.S. deployments, for if the United States sought to deploy defenses today the Soviets could outpace U.S. efforts and leave the United States at a net disadvantage for at least several years.

(Although this is not, in our view, a valid argument for U.S. inaction, it is a factor that the U.S. government cannot ignore in its planning.)

The existence of these extensive, varied Soviet strategic defenses complicates current U.S. targeting plans. Resources must be allocated to defeat Soviet defenses. For example, air defense and ABM sites must be targeted and dispersed forces located; harder silos and command centers require more accurate, higher-yield warheads. Countering the uncertainties posed by Soviet defenses is no small consideration. These are problems that Soviet planners hardly face, for U.S. defenses against ballistic missiles are nonexistent, and defenses against aircraft are nominal. U.S. passive defensive measures are not nearly as extensive. In effect, the Soviets have a free ride to a very substantial portion of the targets they could attack in the event deterrence failed, whereas U.S. offensive forces already have suffered attrition without a shot being fired.

Conflicting U.S. Views

Although the United States and the USSR face different geopolitical problems, the major asymmetry in their defenses is a function of a strategic posture maintained by the United States (since the early 1960s) that focused on deterrence through the threat of retaliation.[6] There has been considerable debate in the U.S. strategic community about what should be targeted in retaliation with strategic offensive forces. Even to many of those who have accepted the need to develop offensive options that would permit U.S. targeting to avoid Soviet cities, the emphasis has been on retaliation. Defense has been seen as largely irrelevant in this debate. The president's speech of March 23, 1983, brought discussion of strategic defense out of the closet in which it had languished for more than a decade. If, as some in the United States have steadfastly maintained, the threat to cities was the principal guarantor of deterrence, asymmetry in the size and composition of U.S. and Soviet forces could be tolerated within very wide margins. It did not matter if Soviet defenses exacted a virtual attrition; the United States would still retain a confident ability to get a significant number of warheads through to Soviet cities, and there was no need for the extensive overkill that was alleged to exist. Of course, if this notion of what should be targeted with U.S. retaliatory capability does not hold up to critical evaluation or does not provide for effective deterrence in a wide range of circumstances, then an asymmetry in offensive forces, potentially complemented by Soviet strategic defenses takes on a very different significance.

Many advocates of mutual assured destruction have contended that deterrence and strategic stability are maintained when the populations and cities of both sides are vulnerable, for neither would dare attack in such circumstances. Defenses are destabilizing, in this view, because they can and will be countered by larger and/or more competent offenses, thus stimulating arms competition. Should defenses really become highly effective (a prospect this school, curiously enough, both denies and fears), the mutual hostage relationship that underwrites stability would be destroyed. This view maintains that the Soviets accepted the inevitability of MAD in agreeing to the 1972 ABM Treaty; the same view also attempts to explain away the massive Soviet defensive program as irrelevant. In this view, the Soviets have wasted their resources on building defenses because the United States can assure destruction of the Soviet urban-industrial complex. In this view, MAD is an inevitable condition and, indeed, considered desirable by some of its advocates. However, many are increasingly reluctant to admit their advocacy, given the Catholic bishops' attack on MAD's immorality and the more recent claim by some scientists that implementing such a strategy could bring on a "nuclear winter." Still, the MAD view dies hard.[7]

Under the MAD strategy, arms control arrangements should assure that defenses do not emerge or, if they do, that they are strictly limited so as to be manifestly ineffective. The solution to U.S. security lies in maintaining the mutual hostage situation and gradually reducing arms primarily through negotiated arms control. We refer to this as the "diptopian" view, as it relies heavily on diplomacy to achieve a utopia in which nuclear arms are frozen and then reduced.

The contrasting view could be called "techtopian." Its goal is mutual assured survival (MAS), a utopia to be achieved by the vigorous application of advanced technology. A number of highly respected U.S. strategic planners for many years have touted the utility of emerging new technology as a means for solving existing and future military problems. Although not all of them agree on the details—not all, for example, believe a perfect defense to be likely—they have tended to brush aside cost considerations, the difficulty of translating research concepts into mass-produced weapons, and, most of all, the political objections raised by European allies to radical changes in strategy.

Techtopians seek a perfect defense against ballistic missiles that would provide a different basis for deterrence and ultimately supplant deterrence based mainly or only on offensive nuclear threats. Many of this school also support development of advanced conventional munitions to assume gradually the role of nuclear weapons. This combination of defense against offensive nuclear weapons, together with the "proxy role" assumed by advanced conventional weapons, would eventually lead to the de facto disappearance of nuclear weapons as an instrument of warfare. Techtopians emphatically reject MAD as an acceptable basis for planning

U.S. strategy, and most are highly skeptical of the benefits arms control can be expected to confer.

A Third View

We suggest that both MAD and MAS are seriously deficient. New technology, whether it be in developing smart conventional weapons or highly effective strategic defenses, should not be rejected as a long-term contributor to a sensible strategic concept, but neither is it a panacea. Diplomacy is an essential complement to a strong defense posture, but it cannot substitute for military power in maintaining deterrence and supporting collective security. This view acknowledges that arms control arrangements, in time, might make a modest contribution to maintaining stability in a world populated by two superpowers and several lesser nuclear powers. However, increased stability does not rely upon arms control as a central contributor to the security of the nation. Rather, this stability should be seen as a potential dividend if equitable and verifiable agreements do in fact materialize.

In this third view, MAS is considered visionary and MAD dangerously inadequate, for MAD has become increasingly self-deterring. It just is not credible that the United States will resort to the massive use of nuclear weapons to respond to threats short of massive use by an adversary, and if MAD is not credible, then it cannot effectively deter a range of serious threats other than attacks on cities. Likewise, it is difficult to imagine that the United States and its allies will create sufficient nonnuclear capabilities to obviate the need to rely, to some degree, on some sorts of nuclear threats in order to deter, at least for the foreseeable future. The reduction of the role of nuclear weapons in U.S. strategy may be desirable, but this goal may not be practical. Such a reduction is not even a clearly desirable objective so long as the United States and its allies are unwilling to raise and maintain a level of conventional forces manifestly capable of deterring Soviet bloc aggression.

It is clear that U.S. allies do not believe reliance on conventional defense, even one based on advanced technology, is in their interest; they have no intention of making more than modest improvements to their conventional forces.[8] The United States, especially under the Reagan administration, has been more willing to meet the demands of a large conventional force—the president's SDI speech contained an often overlooked reference to the development of improved nonnuclear forces. However, it is unlikely that recent levels of defense spending will be sustained by Congress, and even if it were possible to increase outlays by NATO countries on conventional armaments by the planned 3 percent, this would not overcome the significant lead in conventional weaponry enjoyed by the Soviets. It is even less clear that Congress and the parliaments of U.S. allies will support the much larger defense budgets required to supplant nuclear weapons with modern nonnuclear ones. Thus, substantial reliance on nuclear weapons is a fact of life. If, however, nuclear weapons are to continue playing a major role in U.S. strategy, realistic options for their employment must be elaborated. This is, in fact, the direction in which U.S. strategy has been evolving for more than a decade.

The United States has developed a range of options for the employment of nuclear forces that are designed to deny the Soviets confidence in achieving their objectives should they initiate the use of military force. These options seek to permit the measured use of force while minimizing unwanted damage. Their objective is twofold: to deter war in the first instance (which incredible threats to attack cities seem unlikely to accomplish) and should deterrence nevertheless fail, to deter escalation to higher levels of violence while protecting Western interests against Soviet aggression. Those who hold to this concept of deterrence offer no guarantees as to its success. Rather they argue that it is a far more prudent, persuasive way to deter war, and if deterrence fails, it permits a response more consistent with the national interests of the United States and its allies than would reliance on either MAD or a largely, if not purely, conventional defense.

One might gain the impression from some of the recent commentary on these U.S. strategic nuclear concepts that U.S. planners were simply spoiling to fight a nuclear war. Nothing could be further from the truth. These concepts were designed to make deterrence of war more credible by maintaining a plausible capability of responding to Soviet aggression. In fairness, it must be conceded that escalation to such levels of massive nuclear exchanges can never be ruled out with certainty once nuclear weapons are used. But the objective of recent U.S. strategy is to avoid sole reliance on such a threat and to provide the president with options that might successfully protect U.S. and allied interests without the necessity of resorting to the ultimate level of mass destruction.

The Role of Defenses in Deterrence

Defenses can complement a deterrent strategy, even if such defenses are far from perfect, because defenses add uncertainty to the calculations of a planner about the outcome of an attack; uncertainty adds to deterrence. (Of course, defenses add uncertainty to the calculations of both sides.)

Although we have been addressing U.S. planning concepts, from what we know of Soviet planning, it, too, focuses on military targets, not cities. A Soviet military planner contemplating an attack must have confidence that he can destroy a substantial proportion of a diverse, widely dispersed target system, not just a few cities, if he is to accomplish the objective of this attack. Considering the stakes involved, the Soviet planner will presumably be conservative in his assumptions—he will seek high confidence in eliminating the targets he seeks to destroy. If he cannot have such high confidence, he is less likely to attack.

Even defenses that are imperfect, in the sense that they do not protect all possible targets and cannot assure 100 percent protection to those targets they do seek to protect, can add significantly to deterrence. By lowering the confidence of a potential aggressor in the success of its attack, defenses can reduce any rational reason to attack. (By "rational" we mean an estimate of probable success based on assumed postattack outcomes.) With the technology that is now in prospect the United States may be able to create "layered" defenses so that, even if a planner is reasonably confident that some of his attacking weapons will get through one layer, he will not know which ones will penetrate the defense as a whole and where. (This is important if an attacker's purpose is to neutralize a wide range of targets and not simply to hit cities.) The defender may be able to create ambiguously located defenses so that the attacker will be uncertain just what is defended and to what degree. This will create even more substantial uncertainties.

U.S. experience with Soviet defenses is revealing. As was suggested earlier, although defenses have not invalidated the U.S. offensive deterrent threat, they have increased U.S. burdens and have created virtual attrition because a portion of U.S. offensive forces must be allocated to penetrating those defenses. Thus, Soviet defenses have added to U.S. uncertainty in planning the employment of nuclear forces.

Even at the outset, if defenses are designed primarily to protect military targets, they are likely to have some additional capability to protect some portion of the population. In the longer term, as defenses grow, they could limit damage to populations more significantly, even if they were still imperfect. Opponents of defense tend to concentrate their attention on the lives that are bound to be lost if defenses of populations are imperfect, but are they prepared to argue there is no value in being able to save 10, 20, or 30 million lives? Can any U.S. leadership ignore the possibility that deterrence could fail or the prospect of saving tens of millions of lives in a nuclear war, if this can be achieved through deploying defense?

What Is the SDI?

If there is a case to be made for defense, how does the SDI fit into the picture? In some measure that depends upon one's concept of what the SDI encompasses. The Reagan administration has worked to clarify the objectives of the SDI program. Although some ambiguities remain, careful observers should now be able to understand, in general terms, what the SDI is. In particular, one should study carefully Caspar Weinberger's speech of December 19, 1984, to the Foreign Press Center and the white paper issued by the administration on January 3, 1985.

Based on these and other authoritative sources, the long-range goal of the SDI is to create a nationwide defensive shield against ballistic missiles that would help make nuclear weapons "impotent and obsolete." No one can be confident, at this stage, that such an objective can be achieved. If that goal ever were achieved, it would, as the administration has stated, take decades. The technical obstacles are formidable. The administration also has made clear that the SDI has intermediate objectives, although these have yet to be precisely defined. Meanwhile, SDI is a research program designed to explore the feasibility of a range of technologies that could contribute in the future to a defense against ballistic missiles. Incidentally, the program is not designed to develop defenses against airplanes or cruise missiles. To contend that BMD cannot defend against these threats is a truism but also is irrelevant. That BMD would need to be complemented eventually by other defenses if it were to achieve its ambitious long-term goal is obvious. This does not mean, however, that a significant, even if imperfect, defense against ballistic missiles is strategically meaningless.

Much of the SDI debate has been about the possibility of developing a perfect defense of populations. However, the more immediate issues have to do with how the program might develop during the next several years and how defenses that could be created between now and the end of this century—what Secretary of Defense Weinberger has referred to as the transitional stage—will affect strategy, arms control, and U.S.-Soviet relations in general. This depends very much on the precise nature and pace of the program—on what sort of a defense is under consideration. This has yet to be determined in detail, but several possible alternatives can be identified. Which one is chosen will have important consequences for the allocation of defense resources, for arms control, for Soviet reactions to the SDI, for strategic stability, and for U.S. strategy. Any careful analysis should begin by identifying these alternatives.

A minimum SDI program would be restricted to research only, at least until that future point (probably several decades hence) when all the critical components have been tested and promise of a nationwide population defense seems feasible. This would limit the fiscal impact of the program for many years and could be designed to avoid any near-term conflict with the ABM Treaty. (Indeed, as the Reagan administration already has pointed out, research cannot be the subject of sensible arms control agreements because it is impossible to verify compliance.) Because there would be no actual deployments, such a program would be consistent with a deterrent that continues to be based on the threat of retaliation by offensive forces. Most serious opponents of the SDI still advocate continued research. However, the potential of that research cannot be demonstrated unless eventually there is some testing of components. It is likely that, without testing, support for the program would wane. How could the U.S. public and Congress be expected to support multibillion dollar programs for many years with no assurance that the research was in fact developing practical solutions—that is, systems that worked when tested? A research program without testing would not take the nation very far toward the objectives of the SDI—namely, to determine whether key technologies are promising enough to warrant further development. Although such a restrictive approach might satisfy some, it will not satisfy the administration's long-term objectives.

A second alternative also would address the long-range goal of developing technologies that could contribute to a perfect or near-perfect defense against ballistic missiles, but this alternative would proceed beyond research to development and testing. However, this program would have to delineate what was and what was not to be permitted. It would not, for example, seek to develop near-term capabilities. Because this alternative focuses the R&D program on the long-range objective of a perfect or near-perfect defense, the program would direct its attention to the most difficult and demanding technical issues. At least at the outset, a number of alternative technologies would be pursued. For example, for the boost-phase component of the system, research would be undertaken on several types of directed- and kinetic-energy weapons. This would require that research be pursued into development and testing as promising technologies are identified. However, testing could be scheduled so as to avoid conflict with the ABM Treaty, at least for several years. However, as R&D progressed there inevitably would be increasing pressures to test, and concerns about potential conflict with the treaty would arise. In any event, any deployment short of the full system designed to protect populations would be deferred in order to maximize the long-range R&D effort.

Another alternative would pursue an aggressive R&D program directed toward long-term goals, but development of systems that could be deployed before the most demanding technologies could be fully developed also would occur. In this option, the long-range objective of a perfect or near-perfect defense is the ultimate goal, but it is balanced with the intermediate objective of partial defenses that could contribute to deterrence in this century. Thus, options for intermediate deployment would not be precluded during the next several years of research, and if the technologies prove promising, deployments could begin in the 1990s. Examples of such intermediate-system capabilities include airborne optical sensors, high-acceleration, ground-based, kinetic-energy kill vehicles for the terminal phase of the system, and space-based sensors and kinetic-energy weapons for the boost-phase component. There are several variants of this option, depending on how much relative effort is devoted to long-range research versus development of intermediate capabilities. This option almost certainly would be more costly than either of the prior ones because production, deployment, research, and testing would be involved. Therefore, there might be pressure to narrow the range of explorable technologies. This option also would involve activities that would conflict with the ABM Treaty sooner than would the second alternative because the philosophy of this option would be to test promising technologies as soon as they appear ready for testing and deploy if tests proved successful.

Finally, there is an alternative that emphasizes very early deployment by the end of the 1980s. In this case, the primary objective would be to deploy already developed capabilities as soon as possible to defend point targets such as missile silos and command and control facilities. Although more sophisticated components would continue to be developed, less-sophisticated technology that is presently or soon to be available would be given priority. The most mature technologies relate to terminal defenses using radars and ground-based interceptors. (Although the administration has touted the SDI as a nonnuclear system, it is likely in this option that nuclear warheads would be required for very early deployments if the system were to be effective.) This could be supplemented by an early boost-phase defense such as advocated by the High Frontier group and others.[9] These technologies already have been developed, and deployments could begin before the end of this decade if a crash effort were mounted. However, such a program would apply at least some technologies that would probably have little utility in meeting the long-term objective and might divert substantial resources from that objective. Whether such programs should be considered on their own merits is a decision that will have to be made; however, there is no evidence that this approach is currently being considered as part of the SDI.

The line between future research, development, and deployment of new and esoteric technologies and the earlier deployment of some elements of esoteric technology in conjunction with existing defense capabilities is fundamentally a function of policy. In the last analysis, it requires a judgment as to how U.S. security interests would most benefit. In our view, it would be prudent to pursue a range of possible options until the technical possibilities of the SDI are better understood and the intentions of the Soviets in the Geneva negotiations can be further tested.

Stability

Surely a central consideration in making the aforementioned judgment is how the development and deployment of an SDI, however explicitly defined, is likely to affect strategic stability. Some commentators have argued that even if it were possible to achieve a perfect or near-perfect defense, the transition to that goal would involve a period of instability. They contend that the Soviets would respond to a U.S. defensive program by increasing their offensive capabilities in order to maintain their deterrent, and that response could create a new, and by definition unstable, round of arms competition. Furthermore, these observers argue, partially effective defenses combined with improving counterforce capabilities would lead to a situation in which a first strike could appear relatively tempting in a crisis. Although partial defenses would be ineffective against a large first strike, they might be effective against a ragged second strike. With limited defenses on each side, the mutual fear of a first strike could be a source of instability in a crisis. Finally, space-based defenses will themselves be vulnerable and thus an inviting target for attack by both sides. Each of these concerns needs to be addressed.

With respect to crisis stability, the effects of defenses on such stability will depend on the size and character of the opposing offenses and defenses. If defenses were of such a size and nature as to add to the attacker's uncertainty of achieving its objectives and thus reduce the likelihood of an attack, stability could be enhanced. This is particularly true if neither first- nor second-strike options look particularly attractive. Although a first-strike outcome may look somewhat better in theoretical calculations than a second-strike outcome, neither is likely to look very attractive, and even a first strike would be fraught with uncertainty if defenses were deployed. If defenses of land-based forces and command and control increase the uncertainty of the outcome of an attack and reduce the pressures for a prompt response on warning, this would be stabilizing, not destabilizing.

The Western debate about the SDI has focused upon the purported destabilizing effect of U.S. defensive deployments; this debate ignores the fact that the Soviets are, in most respects, in the lead in defenses. Indeed, U.S. intelligence officials suggest that there is evidence that the Soviets may be readying a nationwide BMD deployment.[10] Such a deployment would create problems for U.S. defense planners, but it is by no means clear that this deployment will increase the risk of war. Much will depend upon pace, effectiveness, size, and other aspects of the Soviet program. But, if there were a problem of instability growing out of defenses, the proper concern, in the first instance, should be directed at the Soviet Union.

As to arms race stability, a situation of intense competition exists today. This is primarily a function of the adversarial political relationship between the United States and the Soviet Union. This relationship is not likely to change in the foreseeable future, much as we might wish it would. Thus, one must ask how defenses might alter competition, not whether defenses will create competition. We already have noted that the Soviets are pursuing a wide variety of offensive and defensive programs. It appears most unrealistic to expect them to abandon these programs. Based on the past record, whether or not the United States proceeds with defense, the Soviets, if they find it in their interest, will do so. Furthermore, we expect the Soviets would attempt to counter a U.S. BMD program, in part by modifications in their offensive forces. Just what form these responses will take or how effective they will be depends on a wide variety of factors, including the effectiveness of U.S. defenses.

One often ignored factor is that neither the USSR nor the United States is ever wholly unconstrained in what it can produce, either on its own initiative or in response to the other's action. Given the magnitude of recent Soviet strategic programs, it will not be easy for the Soviets to offset U.S. defenses with new or accelerated offensive programs. Limitations of technology, resources, scientific skill, and production capacity affect what the Soviets can be expected to do. (It affects the United States as well, of course.) We do not yet know whether the Soviets could offset U.S. defenses more easily and cost effectively than the United States can deploy added increments of defense.[11] We do not know whether the Soviets can be induced to follow less destabilizing paths. Surely their massive offensive buildup of recent years could hardly be typified as stabilizing. Is it stabilizing for the Soviets to continue building defenses, which increasingly add uncertainty to U.S. targeting plans, while the United States does little or nothing to defend its own country or its allies?

As to the vulnerability of space-based defenses, this must be acknowledged as a serious problem. Yet very little effort has been expended in the past by the scientific and technical community to determine how the vulnerability of space-based defenses could be reduced. A major objective of the SDI is to explore how space-based systems might be defended. Until this research is carried further, we cannot be certain of the possibilities of competent defenses. However, recent reports do suggest that the United States is in the later developmental stages of a program designed to harden such assets, permit them to maneuver out of harm's way, and defend themselves.[12] Other techniques designed to assure survival are likely to be discovered and employed as the scientific community turns its mind to the task.

Extended Deterrence

Another concern about defense is its impact on extended deterrence—that is, on the U.S. ability to continue extending its nuclear capabilities to help deter attacks against U.S. allies. To place the issue in proper perspective it is perhaps worth restating the obvious: The United States has extended such commitments because it is in the national interest to do so. Successful attacks against major U.S. allies (Western Europe or Japan) could change the world balance of power in a manner intolerable to the United States. Indeed, threats that achieved that end, without actual hostilities, would be no less contrary to U.S. interests. Nor is the spread of nuclear capabilities to states not now possessing such capabilities the best means of assuring allied security.

This explains the need to extend the U.S. deterrent umbrella. But does it suggest that the SDI will be welcomed by U.S. allies as consistent with the fulfillment of past commitments? Apparently not. European allies have expressed intense concerns about and in some cases strong opposition to the SDI. Viewed from the European perspective some of these concerns are understandable, perhaps even valid, but others are misdirected. Some concerns persist because the Reagan administration has done an inadequate job of explaining the SDI and alleviating the fears generated abroad by the program. In particular, the administration initially failed to emphasize the deterrent role of the SDI and the extensive Soviet efforts in strategic defense. At the same time, the United States placed unnecessary pressures on its allies to join in the U.S. effort. Given Soviet efforts, however, it is entirely inappropriate to describe the SDI as an initiative that somehow has forced a Soviet response.

European concerns about the effect defenses may have on deterrence are more complex. In the view of some, NATO allies have placed undue reliance on a nuclear response to Soviet aggression. Surely such a response is not suitable to a range of Soviet military actions, as indeed the flexible response strategy recognizes. Moreover, many on both sides of the Atlantic have failed so far to grasp an essential point—the introduction of defenses must inevitably complicate the calculus of a possible Soviet aggressive action and, as such, inevitably enhances, not detracts from, deterrence.

Many Europeans are concerned with President Reagan's references to the immoral nature of deterrence. Here it is important that the issue be considered in proper perspective. The president and his principal national security associates understand that for many years it will be necessary to rely on the threat of offensive nuclear retaliation. Several of the administration's new programs—the B1-B, the Stealth bomber, cruise missiles, the MX, and others—testify to this fact. What the president has in mind is the immorality of continuing to depend, in the long term, on the threat to kill millions of civilians as a basis for deterrence without a serious effort to seek an alternative. Surely that is not a principle with which Europeans would disagree.

If, as was earlier suggested, the Soviets are placing great emphasis on defense, and if no matter what the United States does or does not do the Soviets will deploy defenses if they consider it in their interest to do so, then allied concerns about Soviet defenses are well founded. Even partial Soviet defenses will tend to diminish the utility of the limited British and French nuclear forces. Moreover, Soviet defenses will tend to make U.S. limited nuclear options more difficult to implement effectively, and these have been developed precisely to make extended deterrence more credible. NATO allies have long had difficulty with the limited options concept, fearing it could lead to a war limited to Europe, but they have gradually come to accept that if a massive nuclear attack on the USSR is increasingly unlikely except in response to a massive nuclear attack against the United States, even less credible would be the massive use of nuclear weapons in response to attacks against U.S. allies. Thus, as indicated earlier, more limited nuclear responses have the virtue of being more credible as a response to such aggression. Even limited defenses reduce the credibility of such a strategy.

Beyond that, however, should there ever be a time where there were highly capable, widely deployed defenses on both sides, this could reduce the willingness of either side to employ nuclear weapons in an attack. U.S. allies are expressing an understandable concern that this could "decouple" the United States from the defense of Europe and make the world "safe" for conventional war—where the Soviets presently have superiority. But this world is in the distant future; the West will have opportunities for addressing this European concern long before defenses reach this level of proficiency. Indeed, if, as many experts believe, defenses never achieve such a level of proficiency, European concerns will prove academic, and the introduction of partial defenses could enhance, not decrease, deterrence. A defense of the United States could add to the credibility of a U.S. response to aggression against an ally or at least increase Soviet uncertainty as to the U.S. response. Surely, if the United States is better safeguarded from a Soviet attack, to that extent the U.S. willingness to respond to Soviet aggression against an ally should be enhanced.

Furthermore, limited defenses against Soviet ballistic missiles designed for theater use could have the same deterrent effect in the theater as in the strategic balance by adding uncertainty to the attacker's plans. NATO is currently deploying a new air defense system (PATRIOT) in Europe to defend key facilities against aircraft attack. That defense is designed to raise the price and uncertainty of an air assault against NATO bases. At the same time, the Soviets are deploying thousands of theater ballistic missiles that pose a threat to the same targets.[13] Unless NATO has a defense against this new threat the Soviets may be tempted to believe that they could disarm NATO in an initial theater missile strike. This could add to instability and to the Soviet ability to coerce NATO in a crisis. Thus, if it proves feasible to develop a defense against theater ballistic missiles as well as against Soviet theater bombers, this could strengthen deterrence and stability in the theater by raising the cost and uncertainty of an attack. The ability of the United States to offer its allies a defense against theater ballistic missiles may help to allay current concerns that defenses would only protect the superpowers and leave other nations more vulnerable.

In sum, although we can appreciate the allied concern (some of it stems from U.S. ineptness in handling of the SDI consultation), the issues are more complex than they have been portrayed to date. A good deal more quiet, less emotional dialogue would be useful. The United States must deal directly with real allied concerns, emphasizing how partial defense could strengthen deterrence and how unilateral Soviet exploitation of defenses can endanger allied security.

The SDI and Arms Control

The impact of the SDI on arms control depends on how one views the role of nuclear weapons and arms control in the U.S. national security strategy. If one's view of a proper nuclear strategy is based primarily upon attacks against urban-industrial targets, defenses—unless perfect—have little strategic value. Even if defenses were perfect, they ought not to be deployed in order to retain the mutual hostage relationship.

A major objective of arms control, in this view, would be to limit defenses. However, if one views the role of nuclear weapons as broader and believes defenses can contribute to deterrence and perhaps ultimately to an entirely new form of deterrence, the relationship between the SDI and arms control will be viewed differently.

As the United States pursues the current arms control negotiations with the Soviet Union, which will include defenses in some way, the United States could have three possible broad objectives. It might seek to reaffirm the ABM Treaty, which already includes substantial constraints on testing and deployment of new types of ABM systems. Such a course would permit continued research on the SDI and some testing, but the SDI program (especially if it included testing) would run up against the treaty—perhaps quite shortly. This will depend on which of the previously discussed R&D options is followed. If the United States seeks to preserve intermediate deployment options, the conflict with the ABM Treaty might occur within a few years. Even if the United States pursues a long-term R&D program, tests will be required to determine the feasibility of various technologies, and at least some of these tests will conflict with the treaty.

A second broad option would be to strengthen and extend the treaty provisions. It is theoretically conceivable that some provisions regarding research could be added to the current treaty, but the verification problems appear insurmountable, and this would certainly be contrary to the current U.S. intention to preserve the SDI research program. The Soviet Union appears to want to extend the treaty (or negotiate a separate treaty) to cover testing and deployment of space-based weapons. However, space-based ABMs already are covered by the treaty. The treaty's main loophole applies to antisatellite systems (ASAT); either side could test certain ABM components as ASATs. Indeed, an ASAT may have certain inherent ABM capability whether or not it is tested in this fashion.

There is another loophole in the treaty that the Soviets do not mention—antitactical ballistic missiles (ATBM). The Soviets do not mention this loophole because they are now seeking to exploit it by developing and deploying antiaircraft and antitactical or antitheater ballistic missile systems that also contain some capability against strategic ballistic missiles. The United States will have to deal with these issues carefully. On the one hand, it hardly seems equitable to try closing a loophole (ASAT) that concerns the Soviets without also closing one that they are exploiting. On the other hand, the United States may want to develop and test an ATBM in the future. If the United States were to try expanding the treaty to encompass ATBM while pursuing the SDI, its allies surely would see this as confirming their fears that the SDI was discriminatory. Thus, extending treaty constraints appears to be a course fraught with perils.

Finally, the United States could seek to relax the treaty through negotiation in order to permit the SDI to be pursued at least into the test stage. There still might be constraints on certain types of ABM deployment, for deployment of many types of ABM is still many years in the future. Nevertheless, this would seem to be the appropriate course to pursue in negotiations if the United States is serious about the SDI.

At this point, a more fundamental question needs to be posed: Should the United States reconfirm the existing treaty or work to strengthen it? Surely, the United States should take either step only if it is in the nation's security interest to do so. We have suggested that defenses could enhance U.S. security even if both sides deploy them. We also have suggested that the Soviets have made great strides in defense under existing treaty constraints.

The United States and its allies need to be realistic about the benefits of a treaty that has been in force for fourteen years. Conditions have changed since the treaty was negotiated. U.S. strategy and targeting doctrine has evolved; mutual vulnerability is no longer viewed as a desideratum of U.S. strategy. Technology also has changed, thereby opening new possibilities for defense. There are indications that the Soviets are now creeping, if not breaking, out from under the current treaty constraints. It is widely acknowledged that the Soviets have violated some provisions of the treaty, particularly by deploying a large phased-array radar in a manner explicitly prohibited by the treaty.[14] Although the treaty presently may serve to constrain more overt, blatant Soviet actions, it seems to have served as a far greater constraint on the United States.

Under the circumstances, it would be a grave error to adhere to an agreement that may no longer serve the national interest. We do not argue that past agreements should automatically be cast aside provided they manifestly serve the national interest.[15] However, we believe the burden of proof rests at least as much with those who would perpetuate past agreements; they must demonstrate that the ABM Treaty, already battered by questionable Soviet compliance practices and by the advance of new technology, should be given priority in U.S. security planning before new defensive technology that could protect the nation and its deterrent forces, in whole or in part.

If the administration is serious about pursuing the SDI—as it seems to be and as it should be—then it must face up to the need to modify the ABM Treaty sooner or later. The treaty need not necessarily be abandoned if the Soviets are prepared to negotiate changes, but it will need to be altered at some point. In fact, the treaty itself includes provisions that explicitly anticipate the possible need for renegotiation. In other words, fifteen years ago U.S. negotiators understood that changing circumstances could require a change in the terms of the treaty. This will be a difficult negotiation but by no means impossible if the United States and the USSR have incentives to negotiate. Continuation of the SDI program would constitute one such incentive.

Notes

1. This view is reflected in the 1983 Hoffman Panel Study prepared for the Department of Defense. See, Future Security Strategy Study, *Ballistic Missile Defense and U.S. National Security: Summary Report* (Washington, D.C.: DOD, October 1983).

2. McGeorge Bundy et al., "The President's Choice: Star Wars or Arms Control," *Foreign Affairs* 63 (Winter 1984/85):265–278.

3. *Wall Street Journal*, January 2, 1985.

4. *New York Times*, February 25, 1985.

5. U.S. Department of Defense (DOD), *Soviet Military Power* (Washington, D.C.: GPO, 1985).

6. For a description of the evolution of U.S. strategy in the 1970s, see L. Sloss and M. D. Millot, "U.S. Nuclear Strategy in Evolution," *Strategic Review* 12, no. 1 (Winter, 1984):19–28.

7. See, for example, Spurgeon M. Keeny, Jr., and Wolfgang K. H. Panofsky, "MAD versus NUTS: The Mutual Hostage Relationship of the Superpowers," *Foreign Affairs* 60 (Winter 1981/82):287–304; and Robert S. McNamara, "The Military Role of Nuclear Weapons," *Foreign Affairs* 62 (Fall 1983):59–80.

8. NATO heads of government agreed in 1977 to a long-term defense program to strengthen NATO forces, including an annual increase in real defense expenditures "in the region of 3 percent." Five countries achieved this goal in 1979 and six in 1980, but only the United States and Luxembourg achieved this goal in both years. By 1983 non-U.S. defense outlays in NATO had declined to approximately 1 percent. See, Roger L. L. Facer, *Conventional Forces and the NATO Strategy of Flexible Response: Issues and Approaches*, R-3209-FF (Santa Monica, Calif.: Rand Corp., 1985), pp. 44–45.

9. For an overview of the High Frontier proposal, see Daniel O. Graham, *High Frontier: A New National Strategy* (Washington, D.C.: High Frontier, 1982). Also see the proposal of Zbigniew Brzezinski, Robert Jastrow, and Max M. Kampelmen, "Defense in Space Is Not 'Star Wars,'" *New York Times Magazine* January 27, 1985.

10. See the testimony of Robert M. Gates and Lawrence K. Gershwin before a joint session of the Senate Armed Services Strategic and Theater Nuclear Forces Subcommittee and the Senate Appropriations Defense Subcommittee, June 26, 1985, (mimeo), pp. 5–6.

11. See the discussions in Gary L. Guertner, "What Is Proof?" *Foreign Policy* 59 (Summer 1985):73–84; and Thomas Krebs, "Moscow's Many Problems in

Countering a U.S. Strategic Defense System," *Heritage Foundation Backgrounder*, no. 454 (September 17, 1985).

12. See the discussion in U.S. Congress, Office of Technology Assessment, *Anti-Satellite Weapons: Countermeasures and Arms Control* (Washington, D.C.: GPO, September 1985), pp. 76–86.

13. There is a whole family of Soviet theater ballistic missile systems, ranging from the intermediate-range SS-20 to the shorter range (tactical) SS-21, SS-22, and SS-23. See DOD, *Soviet Military Power.*

14. For an official perspective of Soviet compliance practices, see the president's *1984 Report to Congress on Soviet Noncompliance with Arms Control Agreements.* For contrasting (private) views, see Colin S. Gray, "Moscow is Cheating" and Michael Krepon, "Both Sides Are Hedging," in *Foreign Policy* 56 (Fall 1984):141–172.

15. For a detailed exposition of the inherent limitations in negotiating a significant nuclear arms limitation with the USSR that is manifestly in the U.S. interest, see Seymour Weiss, "The Case Against Arms Control," *Commentary* (November 1984).

PART TWO

Point Defenses: Halfway House or Strategic Misdirection?

4

Point Defense: Gateway to the SDI

Jonathan B. Stein

Amidst all the clamor and din surrounding the Strategic Defense Initiative (SDI), some observers have grasped the obvious: that the splendid defense envisaged by President Ronald Reagan and by precious few others probably will never come to pass. This is so for technical and political reasons and is becoming clearer to specialists inside and outside the government. Harsh reality, however, will not dim the plans of those committed to ensuring that some kind of strategic defense system gets funded and built. As the SDI moves forward, then, we should expect to see systems and components for defensive missions in the terminal or late midcourse phases to protect U.S. missile silos and command and control centers. Deterrence therefore will be enhanced, according to point or silo defense enthusiasts, and not transcended.

In fact, the diminution of Star Wars already has begun. Boost-phase intercept, the only robust defense possible, was narrowly rejected by the Senate Armed Services Committee as a "realistic goal" for the SDI program in July 1986. Long-term research into comprehensive population defense should continue, said the committee, but "the major emphasis within the SDI should be dedicated to exploring survivable and cost effective defensive options for enhancing the survivability of U.S. retaliatory forces and command, control, and communications systems."[1] In other words, the president's vision is in trouble.

Although limited and relatively small-scale research continued after the Antiballistic Missile (ABM) Treaty of 1972 was signed and ratified, much progress had been made on a variety of technological fronts by the time President Reagan focused on the issue of ballistic missile defense (BMD). The political climate was also particularly ripe, and it seems safe to say that although the headier business of building a comprehensive and robust space-based defense lay well beyond the horizon, a good deal of the newer technologies looked most appealing for the mission of point defenses. Reagan's speech gave added impetus to new research and development efforts but obscured the now mundane task of defending missiles within the atmosphere. After March 1983 the prospect of defending the entire nation (or, indeed, any country or countries) by destroying Soviet boosters in the early stages of their flight from low earth orbit captured most of the headlines and most scholarly attention.

Why Point Defense Now?

Why does the Reagan administration find it necessary to embark on any kind of defensive program? The answer given by the administration is usually a variant of the following: Deterrence needs fixing—it no longer works to U.S. advantage, nor can it stem the rising tide of Soviet nuclear superiority. This theme has been repeated ad nauseum since the mid-1950s and the onset of the first of many "gaps," later shown to be vastly overestimated or completely mythical.

The school of thought that gave us bomber and missile gaps and now gives us windows of vulnerability and a purported BMD gap are known as "strategic rationalists" or sometimes "military traditionalists." Today they are better known as "nuclear warfighters" or "theorists of victory."[2] This class of analysts refuses to acknowledge that nuclear weapons are categorically different from conventional weapons. There is occasional public reference to this important distinction, but as a body of strategic thought their writings obfuscate or deny the difference. All wars, they argue, are fought by military establishments for very definite political and military objectives. Colin Gray has said on more than one occasion that deterrence based on mutual assured destruction (MAD) is both "absurd and murderous," absurd because its proponents cannot specify a viable strategy of waging or terminating war and murderous because countervalue targeting is held in high esteem. The warfighters' focus, then, has been on the inevitability of deterrence breaking down and breaking down catastrophically at that.

Therefore, given the assumption deterrence will fail at some definite point along the conflict spectrum, the only position that makes sense is to devise ways to cope with that inevitable contingency. The nuclear warfighters' emphasis on preparing specific, credible measures to guide and wage nuclear war leads them to embrace counterforce weapons in the triad, comprehensive BMD and the attendant counterforce targeting doctrine that in their view must be joined to these weapons systems. Only thus, do they argue, can Soviet political-military aims be frustrated, crises or wars averted (or won, if fought), and arms control survive. Many strategic analysts who previously came down on the side of "defense emphasis" rather than "victory" are now comfortably united behind President Reagan's SDI. Where better to start on the overambitious SDI than with hard-site point defenses of ICBMs and launch control centers (LCCs)?

Now that the warfighters have thoroughly conventionalized nuclear strategy, they assert that deterrence credibility has eroded since the advent of heavily MIRVed Soviet intercontinental ballistic missiles (ICBMs); they also argue somewhat tardily, that the U.S. fixed land-based missile force became vulnerable to a theoretical first strike (these were the same people who favored MIRVing the U.S. ICBM force to cement a longlasting technological lead!). ICBMs, they argue, provide the backbone of U.S. strategic forces because of their inherent flexibility in targeting, accuracy, reliability, and yield. Soviet predominance in these "prompt, hard-target killers" presumably grants the Kremlin political dividends (blackmail?) and escalation dominance in both a central exchange and in the European theater as well.

The existence of submarine-launched ballistic missiles (SLBMs) and heavy bomber weapons is conveniently factored out of the foregoing analysis as "nondeterring," apparently because these weapons cannot as yet perform prompt counterforce missions with the same confidence that ICBMs can. When future technologies, however, allow the Strategic Air Command (SAC) to reduce the radar signatures (Stealth) on a variety of air-breathing and submarine platforms and weapons, then warning time will be extremely short and much more important than flight time. Sufficient yield, accuracy, and absence of warning time will one day make these countervalue weapons as prompt a counterforce threat as ICBMs are today. Yet the warfighters consistently refuse to acknowledge the march of offensive technologies (or, more precisely, the march of technologies that may have either offensive or defensive applications) in their zeal to promote strategic defense. When these arguments do surface, the warfighters claim that insecure communications between land and sea or the pace of antisubmarine warfare (ASW) technology reduces the value of at least the emerging sea-based counterforce fleet of Trident IIs. But if we are forced to assume that defensive technologies will one day grant us all manner of military opportunities, then we must also assume that command, control, communication, and intelligence (C^3I) problems can be overcome or that the oceans can be made more rather than less opaque.

These rather difficult assumptions are then followed by the now-typical partisan attack on the root cause of U.S. strategic discomfort: arms control and arms controllers, at least as traditionally practiced since SALT I. The crown jewel of arms control, the ABM Treaty of 1972, has of late come in for special scorn and derision owing to its failed promise in the minds of treaty opponents. The treaty promised a leveling off in the growth of offensive nuclear forces and promised both crisis and arms race stability. Critics have long asserted that none of this has come to pass; instead, the SALT process resulted in steady force accumulations centered around the proliferation of MIRVed systems (the Soviets have quadrupled and the United States has tripled its strategic force loadings since SALT I began more than fifteen years ago). It is ironic that today one of SDI's avowed goals is the de-MIRVing of the heavy Soviet ICBMs, but the warfighters insisted on forging ahead with the multiple, independently targeted, reentry vehicle (MIRV) even after severe ABM restrictions obviated the MIRV's utility in penetrating the militarily insignificant Soviet defenses. When MIRVs no longer were necessary to shore up deterrence of a purely countervalue nature, they took on increasing importance as instruments of counterforce doctrine.

The Reagan administration found arms control a tranquilizer for the U.S. people and a blatant gift to the Soviets and now views strategic defenses as the necessary corrective. Defenses can do what the Reagan administration asserts traditional arms control has not done: bolster crisis stability through protection of U.S. retaliatory assets and, eventually, U.S. urban-industrial centers, thereby diminishing the utility of the ICBM as a first-strike weapon. Moscow certainly will see the light, cease producing ICBMs, or at the very least agree to negotiate most or all of them away in the new world of defensive dominance.

Unfortunately, although the ABM Treaty did indeed look to a world of fewer and fewer offensive nuclear forces, the treaty was simply a part of a larger interim agreement that only limited the growth of launchers and not warheads. Therefore, it was inevitable that warhead growth (as much a U.S. concern as a Soviet) would continue unabated and that nuclear force totals on ballistic missiles would expand greatly. Still, the relevant issue is not what the superpowers have done, but what they easily might have done in the absence of any arms limitations agreements. Many additional warheads could readily have been added to both arsenals, especially the Soviet arsenal with its larger missiles, greater throw weight, and the Soviet military's political freedom to add extra launchers and reentry vehicles (RVs) without constraint. In addition, had any sizable defensive program been undertaken by either or both adversary, the launcher/warhead requirements would have steeply climbed throughout the 1970s and early 1980s.

Enhancing Deterrence Through Point Defense: A Dubious Proposition

How exactly would point defense enhance deterrence? As in most other human endeavors, point defense enthusiasts reply, like the ancient Chinese, that the longest journey must begin with the smallest step. Point defense provides that essential first step. Proponents claim that by raising the Soviet attack price—the number of RVs required to destroy a given target—the cost of mounting a concerted counterforce attack will deprive the Soviets of far too many RVs for uncertain gains. Those gains are held to be uncertain because the attacker will not know in advance which RVs will penetrate to the target and successfully destroy it. Being conservative, the attacker either will be forced to expand by several factors the number of offensive forces needed to conduct the attack—which, at some point, becomes prohibitively expensive even if the attacker is not limited before that crunch point by arms control accords—or the attacker will simply, being reasonable, call off the whole thing because the odds of success have grown too complicated and become too uncertain to take preemptive risks in the first place. Confidence in a counterforce strike will have been diminished and the contemplated disarming first strike abandoned in thought or practice.

The theory enunciated here was given full credence by a plethora of reports, analyses, books, and monographs spawned by the Star Wars speech by those who embraced the new emphasis on the "defensive transition." None took the president's vision to heart more earnestly than did the members of the Defensive Technologies Study Team, which prepared during the summer of 1983 technical and political rationales for undertaking a full-blown SDI program.[3] The Hoffman Panel, responsible for policy analysis and particularly interested in examining "intermediate defensive options" and partial missile defenses, concluded that technology "demonstrations can provide a basis for choosing specific, partial missile defense systems *to be deployed by the early 1990s*" [emphasis added].[4] A limited system, in this view, would defend against smaller attacks and be capable of deterring larger, well-planned attacks because of lingering uncertainties on the part of the aggressor. Counterforce arsenals then would fall into disuse and could be more readily cut back or phased out.

Defensive tactics probably would involve the following: adaptive preferential point defenses, wherein the random defense of individual silos and LCCs gives greater leverage to the defense and by definition forces up the attack price because all silos must be targeted, presumably with several RVs. The adaptive technologies, including a variety of infrared sensors and long wavelength frequencies aboard satellites, ground terminals, and the Airborne Optical System, would attempt to discern real RVs from decoys as they enter the atmosphere upon reentry. Interceptors could then be fired or silos protected through so-called simple/novel defenses (such as the "bed of nails" or the "pebble curtain"), thereby deflecting or destroying enough RVs to save the desired portion of the ICBM force to preserve deterrence.

The primary strategic implication of the foregoing argument is that deterrence is premised first and foremost on U.S. ICBMs and therefore that the other legs of the triad do not deter the Soviets in any appreciable manner. Otherwise, why provide one leg with defenses when roughly 60 percent of U.S. strategic warheads are already well hidden (and accordingly protected) on the U.S. submarine fleet? The flip side of this received wisdom is the pat belief that Soviet military planners believe in the validity of limited nuclear warfighting. If not, why posit that they seriously plan to execute a controlled countersilo exchange for as yet unclear or highly ambiguous political objectives?

It is doubtful in the extreme that Soviet Strategic Rocket Forces Command has that much confidence in its technical proficiency in coordinating any plausible first strike. Even if the command did, it is that much more inconceivable that the "cautious, conservative" Soviet elite (characterized as such by almost all of the serious defense hawks) would take such unabashedly foolish, politically devoid risks as undertaking a controlled counterforce strike to settle a diplomatic (or for that matter almost any military) dispute. Not only would such an isolated attack carry an unacceptably high probability of failure on its own terms, but the USSR would then have to confront the majority of U.S. warheads in a retaliatory strike (those undestroyed ICBM RVs and those nondeterring RVs carried aboard U.S. subs and bombers).

Finally, point defenses are said to be legal under the provisions of the ABM Treaty. The treaty's intent, the warfighters argue, is the prevention of a base for a nationwide or territorial defense of the homeland, as spelled out in the preamble and in Article I. (This line of attack is quickly dropped, however, when the SDI's goal is acknowledged to be a comprehensive, nationwide defense.) The argument loses all validity, however, when Articles III, V, and VI and Agreed Statements B, D, and E are considered. No more than 100 defensive launchers, radars, or interceptors are allowed by treaty; the United States elected to defend only one site (the Grand Forks, ND, ICBM base); development and testing restrictions on such systems and components are most severe. Substantial modification or amendment first would be necessary to deploy point defenses, so much so that it is likely that the entire treaty would have to be scrapped and renegotiated.

The Case Against Point Defenses

There are certain very limited, tightly constrained conditions under which point defenses might contribute to stable deterrence. RV totals on ballistic missiles first would have to be severely reduced and maintained at permanently low levels. MIRVed systems would have to be all but eliminated. The cost-exchange ratio between adding extra RVs and extra defensive interceptors would have to greatly favor the defense. A real threat must be presumed that is capable of amelioration; if two-thirds of present or future U.S. forces were to survive any first-strike attack, then the need for such defenses appears increasingly uneconomic—even if all legs of the triad are not evenly weighted.

This is abstract theory. In the real world, these conditions could not possibly hold for very long. The military logic of even limited point defenses leads to expansion of offensive forces, just as similar steps away from the earlier construct of minimum or finite deterrence (steps such as NSDM-242, PD-59, and NSDD-13) required additional refinements away from mutual assured destruction (MAD) and have led directly to the pursuit of active defenses.

The logic of offensive force expansion remains valid because, in the nuclear age, there are only two stable "equilibrium points": mutual assured destruction or mutual assured survival (MAS). Any posture in between these points is unstable; policies or doctrines that avow some intermediate goal or strategy "pulls" force posture toward one extreme or another. For point defense, this can only imply an offense-defense competition and a vigorous expansion of the baseline system of hard-site defenses as now envisaged into all modes of strategic defense—space defense, antisubmarine warfare, civil defense, air defense, anti-terrorist defenses. The system demands incremental additions that strengthen each previous layer. As argued previously, any kind of point defense (and by extension, multilayered defense) can work only after significant force reductions have been negotiated.

Without a strong measure of political détente or at least a substantial easing of hostile rhetoric and poor relations, these kinds of negotiations will not bear fruit. In a strategic climate where defenses are being incrementally added to existing offensive forces—now increasing in number and accuracy—justifiable concerns will be raised that the strategic balance is being upset, that a first strike is being planned or contemplated, and that the need for additional nuclear forces demands force building, not arms control. On the other hand, better relations and less mistrust of intentions would have a much better chance of leading to meaningful arms reductions. In such a climate, the need for point defense becomes far less apparent.

In short, by making offensive forces more secure (through the protection of retaliatory assets[5]), the effectiveness of the ICBM force will be multiplied. When both sides develop point defense, each adversary will view the other's forces as doubly threatening. However, we cannot realistically assume that countermeasures will not be developed. In a crisis the force multiplier granted through survivable defenses will make the ICBM force more threatening and more important to take out early in the crisis because of the additional "bonus" and value ICBM forces will have acquired as multiplicative forces. Thus, neither superpower can allow point defenses to function hindered; countermeasures will be sought and deployed.

There are a host of technical objections to point defenses as well, although it is likely that some rudimentary system of point defense could work reasonably well against a very finite "threat cloud" and in the absence of countermeasures. It never will be easy to discriminate between RVs and reflective decoys within the 30–100 second time frame of the terminal phase. The acquisition, track, commit, and destroy sequence will have to be astonishingly precise and rapid. Kinetic-energy (or, one day, directed-energy) weapons will probably be favored because political distaste will in all likelihood prohibit small nuclear weapons from being used in defense. Direct hits must be scored against tens, if not hundreds, of thousands of targets reentering the atmosphere.

To fool the all-important defensive sensors, a host of feasible countermeasures are either with us today or under intensive development or scrutiny. High altitude nuclear bursts will spread electromagnetic pulse (EMP) across large segments of the United States. Low altitude nuclear bursts through salvage-fuzing the RV will spread local EMP effects, thus resulting either in sensor blackout or redout. Space-based components peering down into the atmosphere to aid in terminal discrimination may be directly attacked. Ground-burst nuclear weapons would raise great amounts of dust and debris, thus blocking the path of ground-based laser systems. Chaff, infrared aerosols, robust decoys, and new approaches in jamming and spoofing all promise to greatly complicate, if not doom, the defensive network's mission.

In addition to these technological difficulties, the Soviet strategic and tactical reaction must be considered as well. The arms race reaction will inevitably be the one described previously—expansion of Soviet offensive forces with particular emphasis on the ICBM force. This is so in part because of the peculiar role assigned to the ICBM in Soviet concepts of deterrence (it accounts for roughly three-quarters of Soviet strategic firepower), but in a wider sense the ICBM represents the cornerstone of Soviet military policy in the strategic arena and is as such the linchpin of Soviet strategic culture. It is not simply a quaint outgrowth of the dominance of the ground forces in the Soviet military establishment or the natural conclusion of a long history of army weapons development beginning with catapults and artillery and culminating in ICBMs. It is rather one of the few tangible measures of Soviet state power that sets the USSR apart from all its peers except the United States. Soviet submarine and bomber programs are small and relatively unsophisticated by comparison; even so, it is only now with the deployment of the new generation of SS-24 and SS-25 ICBMs that Soviet missile engineers have mastered the art of solid fuel propulsion—an achievement reached by the United States more than twenty years ago.

Deployment of U.S. point defenses therefore will encourage Soviet hardliners (and others) to reject any warhead reduction proposals that seek to reduce the value of the Soviet heavy ICBM force below a vague but probably identifiable line drawn by the Soviet General Staff and the Strategic Rocket Forces. The incentive, in fact, leads in a direction unwanted by U.S. planners—the Soviets will add more RVs to their ICBM force and build a point defense of their own outside the strict confines of the ABM Treaty. They thus will be capable of defending, in theory, 75 percent of their strategic forces; U.S. point defense will enable a defense of less than 25 percent of U.S. strategic forces. Regardless of the U.S. declaratory policy of not preempting in a crisis, Congress will not enjoy presiding over a strategic development that grants the USSR a "missile defense gap" in the 1990s.

The strategy of defense suppression will allow the Soviets to continue planning a first strike—if we accept for the sake of argument that such planning is taken seriously by the political leadership—by denying the defenses the ability to operate unhampered and with a high degree of confidence. Suppressing the enemy's defenses, to take the U.S.-Soviet example, could involve using precursor waves of depressed-trajectory SLBMs, air-launched cruise missiles (ALCMs), and submarine-launched cruise missiles (SLCMs), and some older ICBMs in an initial attack designed to disrupt and destroy defensive radars, exhaust the U.S. supply of defensive interceptors, and destroy much of the ground-based elements of a hypothetical point defense system.

This precursor Soviet strike could be followed closely by a much heavier ICBM assault, perhaps timed to arrive on target just several minutes after the first wave. "Punching a hole" in U.S. defenses thus would enable Soviet forces to eliminate as much of the U.S. arsenal as is possible today, only with a higher attack price (and even this so-called "price of admission" need not be exorbitant if offensive countermeasures work fairly well). Point defense, then, cannot offer the United States the option of "riding out" an attack and thereby enhancing deterrence; neither will point defense "complicate" a Soviet attack and thereby deter it.

Pressure will build accordingly to expand the baseline system beyond that of providing a margin of safety to some uncertain portion of the U.S. ICBM force. For example, maneuverable reentry vehicles (MARVs) can be added to the launch vehicle instead of ordinary MIRVs (MARVs are able to change direction once they are inside the atmosphere and are identified by defensive radars). MARV technology will enable the offense to fool the defense down to the very last moment; a sensor will never quite know whether the RV will change direction after an "impact prediction point" has been computed for that RV's final descent or by how much. The leverage of preferential point defense therefore is lost. To regain that leverage, or at least to shore up the final line of defense, it will be necessary to add midcourse and boost-phase defenses. This is indeed what the president's vision is all about, but those uncertain or uncomfortable with the broader implications of Star Wars defenses will have difficulty making the argument henceforth that partial defense of missile silos adds to retaliatory deterrence per se.

There will be serious treaty problems that advocates of point defense have yet to acknowledge. Is it possible to build robust, high-confidence point defenses without providing similar coverage to wider areas? The need for mobility and the co-location of missile fields to a number of population centers in both the United States and the USSR renders a pure point defense concept rather unbelievable. The number of radars, launchers, and interceptors would need to expand to more than 100; the systems and components would have to be MIRVed and made automatic, semiautomatic, or rapidly reloadable; and the necessity of developing, testing, and deploying mobile land-based, air-based, and, very possibly, space-based components for defense in the late midcourse and terminal phases would violate Article V's prohibition against such testing and Article I's ban on the defense of individual regions. This latter point does not rule out all silo defense possibilities. However, for most concepts to succeed a fairly ambiguous degree of mobility will be required, thereby providing some capability for defense of wider areas regardless of stated intentions. In all, at least six articles and agreed statements would be violated. Point defense cannot be secured through reference to the legitimacy of its mission or legal admissibility as the ABM Treaty now stands.

Strategic Stability

There is little evidence that the Soviet military as a body, and certainly the Soviet political leadership, accepts the theory of limited nuclear war or believes in the political utility engendered by "controllable" countersilo exchanges. Soviet doctrine is explicit on these points: Wars that escalate to the tactical nuclear level cannot be kept from escalating to full strategic strikes. There is no evidence that the Soviets believe that only ICBMs would engage in a cosmic duel without other nuclear forces coming into play.[6] In the event of a war that promises to become nuclear at some inevitable point, doctrine posits early attacks with forces ranging from conventional to chemical to nuclear, and in a no-holds-barred "shooting war" point defense will be of little avail in European or central-strategic exchanges.

If the Soviets are fully as Clausewitzian as many conservative defense analysts assert, then they will ask themselves what political objectives are furthered or attained by engaging in a countersilo exchange from which they might emerge with a superior postattack ICBM force ratio, but where they still would be subject to the entire fury of the majority of U.S. nuclear forces. Deterrence is a far broader concern for most Soviet analysts than it seems to be for most U.S. analysts, content as many are to rely on statistical models of game theory for measuring strategic advantage. Classical military strategy, both East and West, insists that political goals be matched with military means. Although there has always been some duality of mind between the Communist party and the military on strategic thought, for many years now defense officialdom has played down earlier notions that nuclear war could be won and has played up the importance of parity.

Asymmetries in each side's defense potential—the expected level of defensive effectiveness—are certain to arise. It is doubtful that both countries would erect identical point defense networks that also are capable of handling with equal skill the set of countermeasures designed to overwhelm and spoof the opponent's network. Even if defense potential were equal on both sides, such equality would not be perceived by civilian and military planners who have made a career out of worst-case analysis. The perception of even slight asymmetries in effectiveness and a highly charged political atmosphere guarantee that new RVs will be routinely added.

If the point defense system on either side were configured such that sufficient ambiguity would exist in a crisis as to whether the point defense would include defense of cities, then crisis stability will be seriously eroded as each side fears that the other is preparing to defend critical territory across each landmass. The incentive in this chaotic environment is to launch first, believing that war has become unstoppable. As regards arms race stability, one of the principal benefits of point defense—that it raises the attack price (some would say significantly) per silo—works against the argument that the offensive threat must be finite and stable for the system to work. The tension between these two positions is apparent and makes it hard for critics of point defense to see just how point defense contributes to arms control.

Assuming for the moment that point defense could perform up to design specifications and protect a significant portion of U.S. retaliatory forces, the effect on strategic stability would not necessarily be positive. (It is further assumed that an offense-defense race has not been set off.) The nature of the particular forces being defended must be taken into account as well. We must ask whether some or all of Minuteman II, III, or MX will be defended. If defended ICBMs MIRVed counterforce systems (like MX or the SS-18, SS-19, and SS-24), then these forces could readily be utilized in a first strike. Each side will fear the "ragged retaliation" syndrome, wherein side A goes first, suppresses the enemy's defenses, and severely reduces side B's arsenal. Side B's retaliation could at best be characterized as "ragged," if not depleted.

If the United States persists with point defense, it is obvious that it will quickly become side B. The United States in all probability will deploy only 50 MX ICBMs; Moscow already has 668 SS-18/19s and can top 820 with the SS-24 in the absence of arms control restraints. The USSR will have between 7,000 and 8,000 counterforce RVs to the United States's 2,000, if Minuteman III (500 of which will be MX) is included in those calculations. If, as warfighters believe, the Soviet leadership can be persuaded to act by the power of numerical advantage, then it is not in the U.S. national interest to abandon the ABM Treaty and actively encourage the Soviets to deploy point defenses.

In conclusion, if it is true that the attack price will be increased, then the Soviets will by definition add additional RVs. Any of the current negotiating postures taken by the United States will be rejected, especially those related to substantial reduction of the Soviet heavy missile force. Soviet point defenses will be added as well, with several negative consequences to follow (as outlined previously). The Soviet political and military elite will retain their version of the Single Integrated Operational Plan at the expense of arms control agreements when faced with the new world of defensive deployments; point defense will not be viewed as the end goal for the United States. The USSR will have to prepare for future U.S. defensive increments. As the essential first step on the road to the SDI, point defense must be reassessed by those convinced of its (dubious) merits.

With the exception of some point-defense enthusiasts, the latter-day defense-emphasis strategists have persuaded themselves that deterrence through retaliation is absurd. They are mired in nineteenth-century conceptions of warfare, yet they insist that technology can hand the United States the means to defend itself completely and comprehensively. But Clausewitz, the master of all post–nineteenth-century strategists, wrote that an absolute defense is an absurdity that completely contradicts the idea of war.[7] As long as the new coterie of Star Wars strategists and counterforce enthusiasts insists on seeing nuclear strategy in conventional terms, they would do well to go back and reread the master.

Notes

1. Senate Armed Services Committee, *National Defense Authorization Act for Fiscal Year 1987,* Report no. 99-331 (Washington, D.C.: GPO, July 8, 1986), pp. 181–182.

2. After Colin Gray's articles since the mid-1970s that "victory is possible." Other theorists in this category include William Van Cleave, Fred Ikle, Albert Wohlstetter, and a host of lesser lights best typified by the work of Daniel Graham. As the official justification for MX is to provide a secure "second strike counterforce capability"—surely an even more absurd notion than countervalue destruction—Gray's support for this and other offensive nuclear systems demonstrates either a failure of logic or a misreading of the Soviet target base, Soviet strategy, and the kinds of nuclear weapons necessary to fight a second-strike war. If Moscow destroys the U.S. ICBM force in a premeditated strike, what realistic counterforce targets will the United States hit in response? Those few older ICBMs held in reserve? All conceivable candidates for retaliatory targeting, such as C^3I, airbases/bomber flyout corridors, SAM batteries, and SSBN/SSN pens, do not require prompt, hard-target kill warheads. If these targets are essential for second-strike kill, then the present arsenal of very accurate warheads will do the job quite well.

3. As Richard Garwin commented at a conference on deterrence at the University of Maryland in September 1984, the Reagan speech reminded him of the scene in Alice in Wonderland where the Queen of Hearts says, "First the verdict, then the trial."

4. U.S. Department of Defense, *Defense Against Ballistic Missiles: An Assessment of Technologies and Policy Implications* (Washington, D.C.: U.S. Department of Defense, 1984), p. 8.

5. This assumes, for the moment, that countermeasures will be absent or ineffective.

6. See, for example, David Holloway, *The Soviet Union and the Arms Race* (New Haven, Conn.: Yale University Press, 1983); Robert Jervis, *The Illogic of American Nuclear Strategy* (Ithaca, N.Y.: Cornell University Press, 1984);

and Raymond Garthoff, "Mutual Deterrence, Parity, and Strategic Arms Limitation in Soviet Policy," in Derek Leebaert, ed., *Soviet Military Thinking* (London: George Allen and Unwin, 1981).

7. Carl von Clausewitz, *On War,* edited by Michael Howard and Peter Paret (Princeton, N.J.: Princeton University Press, 1976), pp. 377–392.

5

Protecting Strategic Forces: The Contribution of Point Defense to Crisis Deterrence

David S. Sorenson

For decades now the declaratory nuclear policy of the United States has stressed deterrence through the threat of massive punishment by attacking an aggressor's homeland. Yet the doctrine has few friends willing to vocally support it and many politically diverse opponents. Most recently, President Ronald Reagan, who would have to give the orders for such a punishment, declared the doctrine immoral and called for its replacement in his so-called Star Wars speech of 1983. Now suddenly the concept of active missile defense is attracting supporters, again from throughout the political landscape—although with differing visions of what missile defense should become.

This chapter will examine the possible deterrent effect of one proposed missile defense, known as "point defense," in providing protection to select and critical military targets. In considering the potential contribution to deterrence that point defense may offer, I will compare it to the present declaratory doctrine of deterrence through offensive threat and mutual vulnerability. Each deterrent type will be specifically evaluated in terms of its potential ability to preserve nuclear stability and reduce the incentive for either the United States or the Soviet Union to preempt under crisis. First, however, some definitions and distinctions are necessary.

Definitions of Deterrence

The concept of deterrence can be reduced to several fundamental propositions.

1. If the cost of an attack to alter the status quo exceeds the expected gains from the attack, the aggressor will refrain from attack.
2. The cost factor can be imposed by either *denial*—preventing the attack by raising the cost to exceed the value of the damage done—or by *punishment* against the attacker to exceed the expected value of the attack.
3. If the military cost (expenditure of military assets) is perceived by the attacker to exceed the value of the attack, the attack will be deterred.
4. If the postattack retaliation on the attacker exceeds the perceived gains of the attack, the attack will be deterred.
5. Deterrence is enhanced when the deterring party reduces the certainty of attack success.

The possibility of conceptual confusion exists here, given the usage of these terms. Snyder would consider the first form of deterrence to be defense, which he defines as a means of reducing the capacity for damage by an enemy.[1] In Snyder's view, deterrence discourages attack by raising the attack cost and risk to exceed the attack benefits expected. Thus, the threat of a defense that increases attack cost and uncertainty about expected attack gains can be said to deter if a potential aggressor believes the threat.

The distinction between deterrence by punishment and by denial is sometimes made with respect to weapons categories. Mearsheimer observes that deterrence by punishment is commonly associated with nuclear weapons, while deterrence by denial has been linked to conventional force.[2] Before the advent of nuclear weapons, the distinction was less clear because the same military force performed both deterrence and denial.[3] Nuclear weapons made the distinction sharper when they were used only to threaten offensive attack; now the distinction between some nuclear weapons and conventional force is blurred, given the reduced yield of certain "battlefield" nuclear weapons. Consequently, the distinction between deterrence and defense also is beginning to erode as nuclear weapons increasingly lend themselves to "warfighting" roles as they approach the size in yeld of large conventional weapons. This has been to some degree true at the strategic level; yield reduction and improved accuracy have made the limited counterforce strike more possible given that the collateral damage might be "acceptable" under conditions of high crisis, an argument that will be developed later in this chapter.

The nature of this distinction between defense and deterrence seems to be emphasized more by the United States than by the USSR, at least in the post-1945 period. The origins of U.S. reliance on air doctrine for offensive power projection are well known,[4] as is the evolution of air defense into deterrence by punishment.[5] Despite doctrinal changes, the emphasis on offensive dominance remains strong in the United States. On the other hand, Soviet emphasis on defense and Soviet rejection of a dominant role for air power as a major military offensive doctrine are also well-known.[6] Soviet doctrine seems to differ in that it rejects the distinction between deterrence and defense.[7] The Soviets seem to prefer deterrence by denial (defense) over the type of assured vulnerability preferred by the United States, which relies on mutuality to succeed.[8] Indeed, Soviet choices on military structure, at least until quite recently, have emphasized defense of the home territory over force projection (aircraft carriers, marines, overseas bases, and such). This difference suggests that divergence also may be found in the way both sides view the role of offense and defense at the strategic nuclear level.

Military Capacity and Deterrence—Offensive and Defensive Distinctions

Distinctions between offensive and defensive military capacity are difficult to make. Definitions are problematic, and tautological traps abound. Offensive capacity, for example, cannot be that which permits expansion into the other's territory. Distinctions between forces commonly associated with either offensive or defensive capacity are difficult to elaborate as well, given that capacity distinctions quickly become confusing when military weapons whose characteristics might be best suited for one purpose are used for another. This was true when tanks—generally considered offensive weapons—were used, for example, to support Israel's Bar Lev Line in 1973 or the Maginot Line in 1940. B-52 bombers, again considered offensive weapons, were used in Vietnam to keep Vietnamese Communists away from U.S. positions—a defensive operation. Furthermore, there are some weapons that defy easy categorization. A battleship is normally used to project power (commonly a property of an offensive weapon), yet can also permit the defense of a harbor. The mobility of weapons is usually one aspect of their offensive capacity, but in some cases the defense, to prevent the advantage of an offensive breakout, must be mobile.[9]

Political intentions, as they may be revealed by doctrine, might clarify things, but doctrine also has ambiguity—as in football, the best defense may be a strong offense. The threat or action to take the war to the homeland of the aggressor may be the best case of this confusion. Of course, some aggression for purposes of expansion is carried out under the cover of defense ("We must eliminate them before they attack us"). Thus, Levy reminds us that striking first should not be confused with aggressive policy, given that striking first may simply be a way to minimize the losses of an impending (or perceived-to-be-impending) attack.[10] Of course, the gleaning of intentions can be confusing and in many cases purposely so. Rarely in modern history have aggressors announced their intention to expand against others out of greed or lust; instead, they justify their expansionist efforts in defensive terms. Napoleon's campaigns were certainly offensive in the sense that the emperor was engaged in military campaigns to expand the amount of territory controlled by France, yet Napoleon claimed that he was "constantly engaged in self-defense."[11] Today the Soviets justify their expansion into Afghanistan by claiming that their actions are to keep Soviet border areas free of Chinese and CIA agents. In 1941 Japanese expansion in Asia was defended out of fears (real or imagined) that the United States had plans to dominate the Western Pacific.[12] In the Seven Years War (1756–1763) both the French and British believed they were acting defensively as they attacked each other's positions in North America.[13]

But despite such definitional and conceptual difficulties, offensive and defensive distinctions can be made with respect to the context of a situation. Those weapons systems and doctrines designed to discourage attacks by raising the attack price are defensive if employed by actors whose primary purpose is to discourage such attacks. Offensive systems are those used for power projection outside national territory when the intention is to alter the status quo, for whatever purpose. This leads into the thicket of intentions versus reality, but there are contextual situations where the intentions are very clear. In 1941 Germany was clearly the aggressor and Yugoslavia was the defender; in 1937 Japan was the aggressor against China; and in 1983 Argentina was the aggressor in the South Atlantic.

Distinctions on some military components also can be made. There are military systems whose clear purpose is defensive (castles, walls, coastal guns, landmines, and such). Obviously such things may protect territory after it has been seized by an aggressor—for example, Nazi guns along the Normandy coast, where, again, the context of the time (June 6, 1944, for example) was defensive.

From the above distinctions, offense and defense at the strategic nuclear level can be conceptualized. In general, strategic nuclear systems designed to strike the homeland of a rival power are considered offensive, and measures to intercept such systems are considered defensive. In deterrence terms offensive systems are those designed to attack first or to inflict punishment for an attack. Defensive systems are intended to raise the cost of such an attack; deterrence is achieved by making attack objectives more difficult to attain. A potential aggressor is forced to exhaust its military force beyond acceptable levels.

The Origins of Mutual Offensive Dominance

The development of the inertially guided ballistic missile as a form of military technology gave rise to one of those unusual instances in history when an active defense against a particular weapon was relinquished. Instead, civilian and military assets were offered as virtual hostages to deter attacks or actions considered inimical to national security. This situation evolved in part because of the prevalent belief that defense against a long-range missile attack was almost impossible,[14] and thus a search for new strategies was initiated. Because active defense against ballistic missiles was considered difficult and thus cost ineffective, it was argued that attacks against national territory would have to be deterred instead by the threat of nuclear retaliation to forestall such an attack from being launched in the first place. However, as both the United States and USSR developed means of retaliation, such capacity became indistinguishable from that required for preemption, and thus the United States developed the concept of mutual homeland vulnerability[15]—a concept often labeled mutually assured destruction (MAD).

The origins of mutual homeland vulnerability are instructive. Although not the sole U.S. nuclear option, it has been the primary U.S. doctrine at the declaratory level since 1963. The doctrine requires both the United States and USSR to credibly target population, industrial, and military sites in the rival power. It became declaratory doctrine under Secretary of Defense Robert McNamara (although earlier origins may be found in the Eisenhower administration). McNamara's initial efforts to get the Soviets to accept assured vulnerability were rebuffed, but the United States continued to hope that sooner or later the USSR would accept the principle. When the Soviets agreed to the Antiballistic Missile (ABM) Treaty of 1972, it was assumed in the United States by some analysts that such acceptance had occurred.[16] The veracity of this assumption is questionable; some now conclude that the USSR simply wanted to halt a more advanced U.S. ABM system.[17]

Assured homeland vulnerability was quite rudimentary as a strategy (or, some might argue, as a condition). It simply required both sides to believe that a nuclear attack (or other incident considered extremely threatening to "national interests") would evoke a punishing nuclear strike against civilian and military assets. Whether, for McNamara, the purpose of the attack was for civilian demoralization or for the destruction of warmaking industrial potential is unclear. But the secretary produced estimates of civilian fatalities and industry that could be destroyed in selective nuclear attacks. These levels of damage represented what McNamara claimed was a threshold beyond which deterrence would be ensured if the Soviets believed that such attacks could take place, especially in the absence of defenses to minimize such damage.[18]

The doctrine of assured vulnerability seemed to have a number of positive features, particularly during the period when civilian targets were emphasized at the declaratory level. The goals contained in McNamara's list (approximately 20 percent of the Soviet population and 50 percent of Soviet industry) could be obtained with fewer weapons than would be required for a counterforce doctrine; once these requirements had been reached there would be no need for additional weapons. A kind of built-in balance would occur once the Soviets matched U.S. capacity with their own weapons, and a kind of nuclear stability would emerge even in the absence of formal arms control agreements. McNamara thus seemed to reflect the thinking of Snyder, who related "stability" to the degree of change in the capacity (political or technical) that gives an incentive or ability to strike first.[19] Hence, instability would occur when one side required only a small proportion of its force to strike first and keep the victim's retaliation to "acceptable" levels.

Oddly enough, the intercontinental ballistic missile (ICBM) now considered destabilizing was once seen as a precursor of stability. Its characteristics have changed from its early days as a weapon too inaccurate to hit anything but cities to its present version, a counterforce-capable system.[20] In fact, the changes in ICBM capacity have raised the stability question, given the missile's increased capacity to strike time-urgent military targets. This capacity in turn threatens the kind of stability that Schelling refers to as "the assurance against being caught by surprise, the safety in waiting, the absence of a premium on jumping the gun."[21] But in the late 1940s and early 1950s this was not a problem—technology had not yet challenged the stability that assured vulnerability seemed to provide.

The emphasis on assured vulnerability seemed to follow from Brodie's maxim that the use of nuclear weapons in war was to be avoided as a matter of national policy.[22] The tremendous damage exchange that would have followed from retaliatory strikes seemed so absurd that the entire utility of war itself was called into question. Thus, assured vulnerability remained a part of U.S. declaratory doctrine throughout the 1970s. Nevertheless, this doctrine was on the wane, as demonstrated by PD-59, issued in 1979, which called for the development of limited nuclear war potential.[23]

Defensive options were never developed extensively in the United States; indeed, for the first decade of the nuclear age they were generally ignored. Some stimulus came from army activity in the development of antiaircraft missiles, but these were limited in nature and would be so until they were upgraded to ballistic missile defense in the late 1950s.

The 1960s were marked by a growing interest in missile defense, but the concept was ultimately rejected in the ABM Treaty of 1972 (modified in 1974). The treaty served once again to reaffirm mutual offensive dominance as the primary strategic choice, and it remains so even in the face of a growing chorus of criticism—most recently, including that of President Reagan. Such criticism suggests that some rethinking may be in order. A brief historical perspective on the nature of offense and defense in history may be valuable in evaluating such criticism and providing some different perspectives to a changing issue.

Offense and Defense in Historical Perspective

There may be lessons from the history of military decisions that could shed some light on the question of whether offensive or defensive dominance is more likely to deter or to provoke conflict. It should be made clear at the outset that only indirect evidence is available, but the results at least may be partially instructive.

A summary view of military history suggests that, in general, military leaders preferred offensive campaigns. The classical period was marked by such in the Peloponnesian and the Punic wars and Alexander's war against the Persians. In European medieval times, offensive strategy dominated, and even the castle, that grand symbol of defense, was used by offensive-minded leaders like Edward I of England as a means of holding territory once captured.[24] The offensive tradition was inspired and continued through the twentieth century by the writings of Clausewitz and Jomini to a point where the offensive was the dominate preference for war preparation, given the assumption that "war had become absolute; victory now seemed to rest upon the speedy annihilation of the enemy."[25]

Offense is expensive, however. The vast losses during the Thirty Years War, the English Civil War, the Muscovite expansion from the sixteenth century forward, and the campaigns of Frederick the Great during the seventeenth century, coupled with indecisive miitary operations (particularly with respect to Cromwell and Frederick) suggest that success in such endeavors is ephemeral at best. Napoleon, who relied extensively on the offensive, could do so only by making war less costly through military conscription and almost total civilian mobilization toward war production.[26] Later history indicates that despite the high costs of earlier offensive efforts, the dominance of offensive strategies continued, as such disparate events as Gettysburg, Napoleon's campaigns in Spain and Russia, the Taiping Rebellion, German strategy in both world wars, and the Japanese Pacific campaign indicate.

The point is not whether the offensive campaign led ultimately to triumph—in the foregoing cases it did not. The point is that it was the instrument of choice for national leaders even when its potentially high costs were understood by such leadership prior to calling up the first troop.

Care must be used, of course, in screening history to make a point—inevitably there are counterexamples. However, a long historical record exists of costly offensives that often ended in failure and great bloodshed. Moreover, not all offensive campaigns have been launched by those who lusted after more territory. The offense also has been the chosen instrument in cases where territorial defense is at least the stated objective. Napoleon, the Confederate States of America, the Japanese prior to World War II, and the Israeli campaign in 1967 (in Gaza and the Golan Heights, at least) are examples of this strategy. In other words, history does support the proposition that offensive strategies, even when adopted for self-protection, may well result in war. As Quester notes, "What is wrong about the offensive technology, under conditions of uncertainty, of even moderate misinformation or hostility, is that the attack is encouraged in the manner of 'shoot first and ask questions later.' Offensive operations then get launched even where states have defensive intentions."[27]

Does an emphasis on defense result in conditions less conducive to war? Again the selective evidence considered here is hardly conclusive, but it is at least suggestive. In China, for example, the Great Wall was completed during the Qin Dynasty (221–208 B.C.) and served to protect the celestial civilization until the Mongol invasion of 1279, thereby providing fourteen hundred years of peace. The perimeter defenses of the Roman Empire helped to maintain relative *Pax Romana* until internal decay in the third century led to weakness and invasion. In European medieval times, "on the whole, medieval commanders were so cautious about defense that they offered few pitched battles."[28] Some might argue that Switzerland has enjoyed more than five hundred years of peace because of the nation's stress on defensive military capacity. One may even speculate that Yugoslavian security after the 1948 break with the USSR had something to do with the Yugoslavian homeland defense capacity.

Doctrines for the Nuclear Age

Nuclear doctrine takes on several characteristics, including the emphasis (offensive or defensive) and the degree to which doctrine relies on mutuality—the condition under which both potential belligerents accept the doctrine and structure their forces in accordance with it. Several doctrinal states are possible: (1) mutual offensive emphasis; (2) asymmetry in offensive/defensive capacity—one side maximizes one type of capacity, the other the opposite type; and (3) mutual defensive emphasis.

Mutual Offensive Emphasis

Under mutual offensive emphasis, weapons assets and other targets remain secure only under conditions of mutual restraint brought about by the credible threat of retaliatory punishment. The paradox of this condition is obvious and has been much discussed—the potential attacker has the incentive to develop a means of firing first to prevent retaliation. The deterring party must then anticipate this and attempt to find ways to make nuclear assets secure from preemption, such that retaliation remains possible. Because the roles of attacker and defender are not always well defined (each may perceive the other as the attacker), both sides may find themselves trying desperately to protect against retaliation on the one hand and attack the other side's retaliatory capacity on the other. As Jervis and others have pointed out, mutual efforts to both protect and penetrate through weapons improvement may well stimulate further impressions of mutual first-strike intention, which can result in a highly destabilizing situation.[29]

Asymmetry

The asymmetrical dominance situation may offer less of an incentive to strike first in a crisis than does mutual offensive dominance. At least one of the two adversaries (the offensive dominant) can have some assurance that the defensive dominant party does not pose to it the "use it or lose it" condition of mutual offensive emphasis. Of course, the side preferring the offense could still use its force to threaten or to attack the defense-dominant side. But it must be assumed that the defense-dominant side would maintain some offensive ability to punish the offender and at the same time require a great expenditure of force to beat down an active defense. This situation is at least preferable in terms of stability to the condition of mutual offensive dominance.

Mutual Defensive Emphasis

Under conditions of mutual defensive emphasis, both sides actively seek to protect those potential targets considered most valuable and most vulnerable. The purpose is to erect a defensive barrier sufficient to cause a potential attacker to avoid aggression in the face of what would be an unacceptable depletion of its military assets. There are several ways that defensive emphasis may be implemented, particularly with respect to the scope of defense coverage. The distinction is usually made in terms of area defense—the protection of cities—and point defense, where protection is limited to critical military targets. The rest of this chapter will focus on point defense, particularly as it may contribute to deterrence through the active protection of ICBM sites and their associated communications and control facilities.

It must be assumed that the calculations for a preemptive disarming strike against counterforce targets would evaluate the postattack force ratio. Presumably, ratios of 1 : 1 would be an unacceptable outcome for a strike. In other words, there is little incentive to strike first if one is rendered no better off by the very act of striking first. This is particularly true in light of calculations that estimate that at least two multiple, independently targeted, reentry vehicle (MIRV) warheads would be required to strike an ICBM in a hardened silo (and possibly more in a "superhardened" silo of more than 25,000 pounds per square inch capable), and even this number gives only about 70 percent confidence of target destruction under good conditions. The kill probability rises to 80–90 percent (at least theoretically) with three RVs, which for discussion sake are assumed to be the minimum configuration for attack success. Thus, to attack successfully a hardened silo requires at least three warheads, all other things being equal. But all other things are not equal; if the attacker has a large ICBM advantage it may be worth the investment of a few MIRVs to attack single-warhead sites. But to do this with a ten-RV MIRV beyond a very limited point is irrational.

Regarding the present and projected U.S.-Soviet ICBM balance, it can be argued that the prospects for a disarming first strike by the Soviets are probably higher than prudent deterrence should allow. This is because the Soviet counterforce capacity has exceeded that of the United States by a ratio of more than 2 : 1 in hard-target capable ICBM RVs, which can theoretically provide counterforce strike capacity. Of course, reasonable people can and do disagree on such a point—after all, the fear of the "window of vulnerability" should have been reduced by the 1983 Scowcroft Report. The report confirmed an earlier study outlining the great difficulty involved in targeting manned bombers and ICBMs in a single, coordinated strike and preventing both legs from surviving to retaliate.[30] The situation involving a first strike would involve a gamble of cosmic proportions with disastrous consequences in case of loss. Indeed, "winning" could bring about severe damage to the "winner" even if it were not hit by a single retaliating warhead (unlikely), given the projected environmental effects of nuclear detonations.

But attacks mounted under dire circumstances are hardly new to military calculations. Throughout the centuries, military operations have been undertaken in conditions where the expectation of success was low and the consequences of defeat high. Certainly that was true of

Hannibal's invasion of Rome through the Alps, whose failure led ultimately to the sack of Carthage. It was true of King Charles XII of Sweden, who initiated (against the advice of his own generals) the Battle of Narva in 1700 with a 5 : 1 inferiority ratio in troops against his Russian foe. Indeed, it was true of many who invaded Russia—perhaps only Hitler deluded himself into believing that his effort in conquering the Russian landmass would turn out differently than had past attempts. The Taiping Army's drive to Peking in 1853 was undertaken under great risk and destroyed the movement.[31] The 1944 Allied invasion at Normandy and the Japanese attack against Pearl Harbor also were undertaken in the face of estimates that failure was quite likely.

There are counterexamples. Most recently, Lebow cites three examples where states did not exploit their military advantage over rivals: Germany in several instances prior to 1914, the United States in regard to the USSR in the 1950s and 1960s, and the USSR vis-à-vis China in the late 1960s.[32] Moreover, it is easy to argue that these cases cannot be compared to conflict in the nuclear age—nothing, after all, can match the destruction of nuclear war, particularly if it should escalate beyond a very limited exchange. But throughout history wars have been total, and the participants realized this—all played by the same rules.

In ancient times total war meant that cities were sacked, crops destroyed, livestock killed. There was more than one Carthaginian Peace, as any visitor on an archeological tour of the Middle East can attest. In Indochina the great Hindu kingdoms of Champa and Funan were obliterated without a trace by the Vietnamese in the fifteenth century. In China similar events took place. During the warring states period (550–476 B.C.) battles were fought until the losing state was annihilated, as was the case during the Taiping Rebellion, where estimates of fatalities exceeded 20 million, and some provinces were flooded by human bodies damming up streams. Such consequences were not restricted to that period—reports permeate Chinese history of wars of total annihilation, where no prisoners were taken except for slaves.[33] Similar horror can be seen in Europe during the Thirty Years War, which reduced much of central Europe to a starving, depopulated ruin, especially in Germany, where 30–40 percent of the population may have been lost.[34] The wars of Louis XIV lasted for more than forty years and consumed more than 2.4 million lives, more than did the Thirty Years War.[35] Not only were memories of that previous conflict insufficient to deter Louis XIV, but indeed the Sun King ordered his generals to emulate the destruction of the Thirty Years War.[36] By early 1915 the belligerents in Europe had realized, for the most part, that a long war of attrition lay ahead of them,[37] yet the offensive continued on, even in the face of the staggering 1914 losses—nearly half of French war deaths, 640,000, occurred between

August and December 1914.[38] In 1941 Japanese leaders were well aware of the enormous risks involved in provoking the United States, and they were correct. For Japan the result was total war and vast destruction, including the obliteration of most of the country's major cities.[39] Nor did the prospect of huge losses serve to deter Vietnam's communist leaders from committing their society to victory, and despite the massive losses incurred, there seemed to be no "breaking point" preventing the Vietnamese Communists from reaching their goals.[40] The massive losses suffered by Iran in its war with Iraq (estimated to exceed 500,000 deaths during six years) suggest that the willingness to pay a high price in war continues on to the present. Indeed, one might argue that the prolonged destruction in Lebanon during the past decade might well exceed that of a promptly terminated, limited nuclear war.

Such cases do not prove that a Soviet Union backed to an intractable position would attempt nuclear preemption. Soviet adverturism may be constrained by memories of the horrifying losses suffered in World War II and by the conservative risk-averse style of present Soviet leadership. On the other hand, the USSR, if engaged with the United States in a crisis where mutual suspicion of preemption arises, might test U.S. threats to retaliate. As Jervis suggests, "If the primary goal is deterrence . . . then the impracticality of the doctrine (deterrence by punishment) is more of a problem. If Washington says the Soviets will be deterred only if U.S. forces can perform missions which in fact they cannot—and which Russian leaders think their American counterparts know cannot be carried out—then Soviet adventurism might be encouraged."[41]

Douglass and Hoeber argue that the Soviets not only believe that a nuclear war could be a short war (the kind they are said to prefer) but that they could prevail by catching the United States unprepared in a surprise attack.[42] Lambeth contends that the Soviets are likely to preempt in periods of high crisis and further notes what may be a growing interest in targeting of critical military sites in the United States.[43] Cimbala states that if "looked at from the Soviet end of the barrel, failure to preempt would be politically un-Marxist and militarily self-defeating."[44] Even if such a prospect were unlikely, there are historical precedents for it. Stability, the condition under which the temptation for preemption is discouraged, might be enhanced more by an active defense than by the continued threat of punishment that increasingly has become more risky to the United States and thus less credible as a deterrent.

An adoption of point defense could enhance deterrence, if it were shown that point defense could make significant contributions toward crisis stability and that the system could be given the technical ability to intercept at least a proportion of attacking warheads. There is no inherent reason why stability must be associated with offensive emphasis—in fact, given the arguments advanced here, offensive emphasis with modern accurate missiles may well be destabilizing. Stability may be better enhanced by raising the attack cost to exceed the expected attack gains, if such can be done.

This raises the question of the technical feasibility of point defense with respect to the required level of intercept capacity necessary to deter attacks. Consider the following 1990s-era scenario, where a Soviet strike is directed against U.S. land-based, counterforce-capable systems. These are presently limited to the Minuteman II, now being upgraded to hard-target kill capacity, the Minuteman III (550 launchers), and the MX Peacekeeper force, configured in the following manner:

1. Minuteman II, each with 1 warhead—450 warheads total
2. 550 Minuteman III, each with 3 warheads—1,650 warheads total
3. 50 MX Peacekeepers, each with 10 warheads—500 warheads total

for a total of 2,600 land-based, counterforce systems on 1,050 launchers.

To attack this force, the Soviets could draw from their ICBM stocks, expected by the 1990s to be structured as follows:

1. 150 SS-17s, each with 4 warheads—600 total
2. 308 SS-18s, each with 10 warheads—3,080 total
3. 360 SS-19s, each with 6 warheads—2,160 total
4. 100 SS-24s, each with 10 warheads—1,000 total
5. 400 SS-25s, each with 1 warhead—400 total

for a total warhead count of 7,240 on 1,318 ICBMs (all figures above are taken from the 1985–1986 *Military Balance*). It is generally assumed that 3 warheads per silo are necessary for a .9 kill probability; thus, the Soviets will require 3150 RVs to destroy 2,340 on 945 ICBMs. The Minuteman II system already has cost the Soviets, under this scenario, 1,350 RVs to destroy 450 RVs. The Soviets may well choose not to extend their attack to these sites and instead target only Minuteman IIIs and MXs—a total target count of 2,150 RVs on 600 launchers. So, 1,800 warheads would be invested, probably costing the Soviets around 180 launchers if they use the SS-18 and SS-25 force with 10 MIRVs each. With 215 surviving RVs, the U.S. counterforce capacity against remaining Soviet hard-target missiles would permit the destruction of 72 of them, with, for estimation pruposes, a total of 432 warheads (an average of 6 per ICBM). The Soviets then hold an advantage of 5,008 to 0 in hard-target-capable RV after both attacks; if 6 is used as an average MIRV load on the survivors, 835 Soviet ICBMs remain while U.S. missile silos stand either empty or destroyed.

None of these calculations takes into account the near-term threat to fixed missile silos from counterforce-capable submarine-launched ballistic missile (SLBM) attacks. The United States and USSR are working on new SLBMs and associated space-supported guidance systems that will theoretically be capable of counterforce attacks by the late 1980s. Although obviously it is premature to speculate on force posture, the very nature of this emerging technology should make more urgent the task of point defense. The Soviet ICBM force may be upgraded in the mid-1990s, which suggests that the present force will be almost entirely replaced by a mix of new weapons, including the SS-24 and SS-25 as well as new heavy ICBM replacing the SS-18. It appears that only a small fraction of SS-19s will remain from this decade's force loadings.[45] But such a prospect will not be likely to involve a substantial change in the overall numbers presented above.

Finally, this first-strike scenario does not consider the prospects of U.S. retaliation by SLBMs and manned bombers. However, until those systems are upgraded, they remain incapable of much hard-target capacity. They of course can be used against other soft military (and civilian) targets, but only at the risk of a further escalation of violence. The U.S. willingness to retaliate under such conditions is uncertain, and this alone may be sufficient to degrade deterrence under a crisis situation.

What do these preemptive counterforce results look like when active point defense is introduced? Of course, the efficiency of such a system is a major determinant of its potential deterrent effect, and such efficiency is very difficult to determine, especially under real operating conditions. But a system efficiency of .5 is not unrealistic—indeed, it would probably not be worth the investment for a system capable of less. With the defense provided by a .5 efficient point system the Soviets would have to double their attack to achieve the results above, and thus 3,600 RVs on 360 ICBMs would be needed. The Soviets would still have approximately 3,064 warheads on 510 ICBMs (again an estimated 6 per ICBM). However, at this point the advantage is worth much less given that there are remaining Minuteman IIs, 450 in all, which require another 1,350 RVs to destroy them with .9 confidence, or 225 additional Soviet hard-target-capable ICBMs. At this point all that remains of the original 1,318 hard-target ICBMs are 285 launchers and presumably a large number with only 1 RV. The Soviets, should the conflict escalate, would then face U.S. superiority in the sea-based and bomber legs of the triad, and escalation would no longer run in their favor. This, again, with a .5 efficient system; the results obviously would improve if efficiency were raised to, say, .75, but again the point is that a near-perfect system is not necessary for deterring a counterforce attack.

All of this is, on the one hand, a highly idealized set of calculations, depicting only one of a number of attack scenarios. It does not directly consider attacks against command, control, communication, and intelligence (D^3I) targets, for example, although point defense could be expanded to cover C^3I targets that can neither be hardened against attack nor widely dispersed to complicate attack plans. But the addition of even a relatively low level of defense complicates Soviet counterforce attack scenarios in a way that should enhance deterrence more effectively than may additions of offensive force. In fact, the evidence presented in this chapter suggests that more offensive nuclear force will erode deterrence, given that planned force additions for the United States stress the very counterforce that makes initiation tempting for the USSR—as it has for the United States.

Potential Problems Associated with Point Defense

Any change in doctrine and its associated weapons systems may be marked by possible problem areas, and point defense is no exception. First, if point defense were deployed by the United States, the USSR would simply shift its targeting emphasis to cities and other soft targets, leaving only charcoal for the point-protected U.S. missiles to defend in a postattack environment. But if the Soviets intended to destroy cities and other soft targets, then a blanket, near-perfect defense would be required to protect against an aggressive USSR bent simply upon the destruction of the United States. Or, for that matter, mutual homeland vulnerability, as part of declaratory doctrine, can be used to try deterring such an attack. But in this chapter I have argued that the Soviets will preempt against what they view as most threatening. U.S. cities are not threatening to the Soviets; U.S. counterforce is.

A second, more serious potential problem for point defense involves the degree to which the Soviets come to see the concept as provocative. The Soviets may view point defense simply as a means to protect high-value targets against retaliation, thus giving its possessor a first-strike capacity. However, the degree to which point defense is provocative depends on what points are defended. If the Soviets truly fear U.S. preemption, then they must realize that the forces used to carry out such an action would come largely from land-based U.S. ICBMs, the very force that would be most likely to receive point protection. Then why protect ICBM sites if the United States plans to use them preemptively? There is little value in actively protecting a scorched empty hole. Silos might remain valuable targets only if the United States maintained missile spares that would permit silo reloading; but this is not present policy, and there is no indication that this will change in the foreseeable future.

Even if the United States obtained point protection capacity ahead of the USSR, such a development might not be as provocative to the Soviets as some may believe. Of course, a preferred state would be mutual point protection of time-urgent targets—a probable occurrence in the near future given the Soviet defensive emphasis and the particular need for the Soviets to protect the large proportion of Soviet silo-based strategic assets.

Other problems emerge with respect to arms control. Even a limited pont defense would require more capacity than the present ABM Treaty permits—100 interceptors at only one site. Furthermore, the treaty's radar limits are insufficient to protect the 550 silos (or possibly 1,000 if remaining MMIIs also are to be protected), especially given the geographical dispersion of those silos. The treaty would have to be modified or abrogated, and the Soviets could require a high price for agreeing to modification.[46] But, defense of counterforce might prove attractive to the Soviets themselves because 70 percent of their total yield is in land-based silos. Thus, it is possible that some modifications to the ABM Treaty might be acceptable to both the United States and USSR simply to provide a modest level of point defense.

Any attempt by the United States to modify the ABM Treaty could cause real concern in Europe, especially in those nations that have developed a limited nuclear capacity: Britain and France (and China). It could be argued that modifications of the ABM Treaty to permit point defense would erode the limited deterrent capacity possessed by those nations. But, for the most part, British, French, and Chinese nuclear systems are incapable of hard-target capacity. (The British future D-5 capacity is an exception, should it occur.) Modifications to the treaty to allow point defense would not weaken what is basically a "massive retaliation" deterrence by these other nuclear nations. Moreover, both Britain and France, as NATO members (France inactively), have not eschewed their stated reliance upon U.S. central strategic systems as the final deterrent for the defense of NATO. It might therefore be somewhat reassurring to them to see the most vulnerable leg of the U.S. strategic triad receive some modicum of protection, given NATO's stated reliance upon it for deterrence.

An additional problem, raised by Glaser, is the added uncertainty that mutual defense might create should the Soviets also adopt point protection.[47] The confidence that such a system would work could erode deterrence in a time of crisis. Yet such a problem also can be seen as an asset, given that deterrence is generally enhanced by uncertainty on the part of a potential aggressor that its attack may not succeed.[48] This may be particularly true of the Soviet Union, where risk-averse behavior seems to be characteristic.

Another difficulty arising from point defense is that it may stimulate more effort at the offensive weapons level by the opposition. More ballistic missiles may be built, along with more decoys, chaff, depressed trajectories, and such, or efforts could be refocused on nonballistic cruise missiles. But it is hard to argue that such developments would not occur if offensive-based assured vulnerability continues. In fact, such developments have occurred on both sides of the superpower force structure. It is conceivable that some of the present offensive weapons effort would be transferred to defensive weapons should both sides adopt defensive doctrines, and this would at least lead to a less-destructive arsenal than the present course may.

Defensive efforts would be more productive under an arms control regime, such as SALT II, where some limits do exist on additions to the offensive force. The United States and the USSR may well value the predictability that arms control provides if major shifts in doctrine should occur. Investments in defensive capacity are much safer to make if there is reason to believe that both parties value the condition of a status quo on offensive forces.

Finally, point defense raises the obvious technical issues that will affect how well it will work. Although a comprehensive discussion of technical issues is beyond the scope of this chapter, those who have addressed the technology of point defense argue in general that technical problems must be overcome but that the technical task of point defense is much simpler than is the case for area defense.[49] This is largely due to two factors: the greater concentration of defensive systems around point targets and the lower requirements for overall success. As previously noted, point defense does not require near-perfect capacity, but area defense does, given that even a small amount of leakage in the system woudl cause substantial population losses.

Conclusion

The concept of assured homeland vulnerability, once perhaps a viable deterrent in the multimegaton age, appears to be eroding, because of the slow but steady convergence of two factors that may together threaten deterrence as provided solely through offense-dominant threat. The first factor is the ongoing development and deployment of technology that is changing the character of strategic nuclear weapons from instruments of mass destruction to war-capable weapons. This is largely due to their increase in accuracy and reduction in yield to the kiloton range—indeed, defused megabombs are now displayed in museums. Yet the strategy that was first formulated in the days when these museum pieces were operational remains U.S. declaratory doctrine, albeit with changes. These very changes may make the doctrine even more destabilizing in light of the present emphasis on counterforce targeting and the kind of "prisoner's dilemma" behavior that it may induce during a severe crisis where national goals appear to be seriously threatened.

The second factor is the outer limit to which societies are willing to take casualties and damage in order to preserve their defined political goals. No one knows what those limits might be in a given prebellum state, but historical evidence suggests that the Soviets (like some other contemporary societies) may be willing to risk millions of deaths if there were no perceived alternative to the expected loss of critical Soviet interests. Such potential willingness can no longer be dismissed as a product of the paranoid U.S. right, particularly in the face of the growth in counterforce capacity that seems to be leading in the direction of earth-penetrating warheads, close-to-zero CEPs, and conventionally armed ICBMs. Should U.S.-Soviet relations deteriorate rapidly under crisis conditions, the very nature and capacity of offensive strategic weapons could render a decision to wait for the other side to attack very difficult.

This is not to suggest, however, that offensive strategic nuclear weapons be abandoned along with the doctrine of assured homeland vulnerability. Indeed, assured homeland vulnerability, when applied to threats against population targets, may have contributed to deterrence. But the choice to strike first against cities is far less likely than is the choice to strike against counterforce-capable weapons.

Even the modest improvement that an active defense of counterforce targets could make to the problem might well be worth the effort, despite the costs of doing so. It would not end deterrence based upon the threat of countervalue retaliation, which invokes a level of damage that in the absence of massive civil defense programs would remain unacceptable to even the most irrational leaders. But point defense would move both superpowers at least a short step away from the hair-trigger world of counterforce, and that, in and of itself, is an argument for the serious consideration of point defense.

Notes

An earlier version of this work was presented to the annual convention of the International Studies Association, March 1985. The author wishes to thank Joseph J. Kruzel, William J. Bishop, Carolyn L. Bell, Stephen J. Cimbala, Elizabeth B. Wharton, Charles F. Hermann, and Joel Wit for their constructive advice while acknowledging that none of them will agree with everything in this chapter and cannot be held responsible for it.

1. Glenn H. Snyder, *Deterrence and Defense: Towards a Theory of National Security* (Princeton, N.J.: Princeton University Press, 1961), pp. 1–5.

2. John J. Mearsheimer, *Conventional Deterrence* (Ithaca, N.Y.: Cornell University Press, 1983), p. 15.

3. Snyder, *Deterrence and Defense,* p. 8.

4. Bernard Brodie, "Some Notes on the Evolution of Air Doctrine," *World Politics* 7 (April 1955):349–370.

5. Michael Mandelbaum, *The Nuclear Question* (Cambridge: Cambridge University Press, 1979); and Desmond Ball, *Politics and Force Levels* (Berkeley: University of California Press, 1980).

6. Owen Jensen, "Classical Military Strategy and Ballistic Missile," *Air University Review* 35 (May-June 1984):54–63; and Brodie, "Some Notes on the Evolution of Air Doctrine."

7. Robert Jervis, *The Illogic of American Nuclear Strategy* (Ithaca, N.Y.: Cornell University Press, 1984), p. 42.

8. Ibid., p. 102.

9. Mearsheimer, *Conventional Deterrence,* pp. 25–26.

10. Jack Levy, "The Offensive/Defensive Balance of Military Technology: A Theoretical and Historical Analysis," *International Studies Quarterly* 28 (June 1984):229.

11. Emmanuel Las Cases, *Memoirs of the Life, Exile, and Conversations of the Emperor Napoleon,* vol. 5 (New York: Peter Eckler, n.d.), p. 103.

12. John Toland, *The Rising Sun* (New York: Random House, 1970), ch. 5.

13. Richard Smoke, *War: Controlling Escalation* (Cambridge, Mass.: Harvard University Press, 1979), pp. 199–200.

14. This conclusion was enunciated in the Truman-Attlee-King statement of November 1945, which described "means of destruction hitherto unknown, against which there can be no adequate military defense." See Bernard Brodie, *The Absolute Weapon: Atomic Power and World Order* (New York: Harcourt Brace, 1946), pp. 23–24. Brodie came to the same conclusion when examining the V-1 and V-2 German missile programs and hypothesizing about the effect of atomic armed missiles (pp. 28–32).

15. The term "homeland vulnerability" as used here refers to the vulnerability of both civilian and military targets in the homeland of each side.

16. Mandelbaum, *The Nuclear Question,* p. 211.

17. Robert P. Berman and John C. Baker, *Soviet Strategic Forces: Requirements and Responses* (Washington, D.C.: The Brookings Institution, 1982), p. 62.

18. Lawrence Martin, "The Role of Military Forces in the Nuclear Age," in Lawrence Martin, ed., *Strategic Thought in the Nuclear Age* (Baltimore; Md.: Johns Hopkins University Press, 1979), p. 5.

19. Snyder, *Deterrence and Defense,* p. 97.

20. Lawrence Freedman, *The Evolution of Nuclear Strategy* (New York: St. Martin's Press, 1983), p. 192.

21. Thomas C. Shelling, *Arms and Influence* (New Haven, Conn.: Yale University Press, 1966), p. 235.

22. Brodie, *The Absolute Weapon,* p. 76.

23. Donald M. Snow, *The Nuclear Future: Toward a Strategy of Uncertainty* (Tuscaloosa: University of Alabama Press, 1983), p. 10.

24. B. H. Liddell Hart, *The Decisive Wars in History* (Boston: Little, Brown, 1929), p. 50.

25. Jay Luvaas, "European Military Thought and Doctrine, 1870–1914," in Michael Howard, ed., *The Theory and Practice of War* (Bloomington: Indiana University Press, 1965), pp. 76–77.

26. J. F. C. Fuller, *The Conduct of War, 1789–1961* (New Brunswick, N.J.: Rutgers University Press, 1961), pp. 32–49.

27. George Quester, *Offense and Defense in the International System* (New York: John Wiley and Sons, 1977), p. 11.

28. J. F. Verbruggen, *The Art of War in Western Europe During the Middle Ages* (Amsterdam: North Holland Publishing, 1977), p. 288.

29. Robert Jervis, *Perception and Misperception in International Politics* (Princeton, N.J.: Princeton University Press, 1976), ch. 3.

30. Alton H. Quanbeck and Barry M. Blechman, *Strategic Forces: Issues for the Mid-Seventies* (Washington, D.C.: The Brookings Institution, 1963), p. 78.

31. Yu-wen Jen, *The Taiping Revolutionary Movement* (New Haven, Conn.: Yale University Press, 1973), ch. 9.

32. Richard Ned Lebow, "Windows of Opportunity: Do States Jump Through Them?" *International Security* 9, no. 1 (Summer 1984):150–151.

33. Herbert Franke, "Siege and Defense of Towns in Medieval China," in Frank Kierman and John K. Fairbank, eds., *Chinese Ways in Warfare* (Cambridge, Mass.: Harvard University Press, 1974), p. 186.

34. Theodore K. Rabb, *The Thirty Years War* (Boston: Heath, 1964), p. 46; J. V. Polisensky, *The Thirty Years War* (Berkeley: University of California Press, 1971), ch. 8–9; and C. V. Wedgwood, *The Thirty Years War* (New Haven, Conn.: Yale University Press, 1939), ch. 12.

35. Levy, "The Offensive/Defensive Balance of Military Technology," p. 357n.

36. John B. Wolf, *The Emergence of the Great Powers, 1685–1715* (New York: Harper and Brothers, 1951), p. 42.

37. Liddell Hart, *The Decisive Wars in History,* ch. 4; and Fritz Fischer, *Germany's Aims in the First World War* (London: Chatto and Windus, 1967), ch. 6.

38. William H. McNeill, *The Pursuit of Power* (Chicago: University of Chicago Press, 1982), p. 318n.

39. Toland, *The Rising Sun,* ch. 4–5; and Nobita Ike, *Japan's Decision for War: Records of the 1941 Policy Conferences* (Stanford, Calif.: Stanford University Press, 1967).

40. John Mueller, "The Search for a 'Breaking Point' in Vietnam: The Statistics of a Deadly Quarrel," *International Studies Quarterly* 24, no. 4 (December 1980):497–519.

41. Jervis, *The Illogic of American Nuclear Strategy*, p. 102.
42. Joseph D. Douglass and Amoretta M. Hoeber, *Soviet Strategy for Nuclear War* (Stanford, Calif.: Hoover Institution Press, 1979), p. 12.
43. Benjamin S. Lambeth, "Selective Nuclear Options in Soviet Strategy," in Johan J. Holst and Uwe Nerlich, eds., *Beyond Nuclear Deterrence* (New York: Crane, Russak and Co., 1977), pp. 93–101.
44. Stephen J. Cimbala, "Is a Soviet 'Bolt From the Blue' Impossible?" *Air University Review* 36 (May-June 1985):28.
45. *Aviation Week and Space Technology*, July 29, 1985, p. 70.
46. Robert C. Gray, "Arms Control Implications of Ballistic Missile Defense" (Paper presented to the Joint Section on Military Studies/Strategic Studies Institute Conference, U.S. Army War College, Carlisle, Pa., October 20–22, 1982).
47. Charles L. Glaser, "Why Even Good Defenses May Be Bad," *International Security* 9, no. 2 (Fall 1984):120–121.
48. See, for example, Stanley Sienkiewitz, "Observations on the Impact of Uncertainty in Strategic Analysis," *World Politics* 32, no. 1 (October 1979):90–110.
49. Sidney Drell, Phillip J. Farley, and David Holloway, "Preserving the ABM Treaty: A Critique of the Reagan Strategic Defense Initiative," *International Security* 9, no. 2 (Fall 1984):78; and Ashton B. Carter, "BMD Applications: Performance and Limitations," in Ashton B. Carter and David N. Schwartz, eds., *Ballistic Missile Defense* (Washington, D.C.: The Brookings Institution, 1984), pp. 122–157.

PART THREE

SDI and the Atlantic Alliance: Strategy and Policy Issues

6

SDI: The Transatlantic Challenge

David N. Schwartz

For reasons of history and politics that are familiar to most, strategic defense until recently was relegated to the backwaters of the strategic debate on both sides of the Atlantic. Although the 1972 Antiballistic Missile (ABM) Treaty put an early halt to defensive competition, small, dedicated groups of research scientists and engineers in the United States devoted considerable energy to exploring the technical limits of ballistic missile defense (BMD) within treaty constraints. Several hundred million dollars have been spent each year since 1972 to finance these efforts.[1] Indeed, as the United States tried to address the growing problem of land-based intercontinental ballistic missile (ICBM) vulnerability, many proposed schemes included plans for modest ballistic missile defense deployments to enhance ICBM survivability. But the U.S. program attracted little if any high-level political attention. Within U.S. academic circles it received equally short shrift.

Given the generally low priority accorded to strategic defense within the U.S. national security establishment prior to 1983, it is perhaps not surprising that European governments and academic circles also paid little attention to it. The ongoing U.S. R&D program was certainly no secret, and it seems implausible that Europeans were unaware of it. But given the ABM Treaty, which guaranteed that any new deployments would be severely limited, and a pronounced lack of interest in Washington even for exercising its full rights under the treaty, Europeans could be forgiven for assuming that strategic defense was a dead issue. After all, there were far more pressing items on the alliance's agenda: detente, SALT II, intermediate-range nuclear forces (INF), to name but three. The same reasons that led U.S. academics to discourage their graduate students from pursuing doctoral theses in this area probably operated as effectively at Oxford and Cambridge. The net result was a decade of silence on the subject among the Western European allies.

Consequently, on both sides of the Atlantic a strategic world characterized by robust offensive competition but rigidly controlled defensive developments came to be taken for granted. Only rarely was it pointed out that such a situation, in which dominance of offensive forces was a central characteristic, was hardly a natural or even an inevitable state of affairs. Occasionally someone would point out the historic anomaly of the situation; just as rarely, a maverick strategic analyst would provoke a chorus of protest by suggesting that strategic defense should play a more important role in Western strategy. But the politics and history of the issue conspired to make it virtually impossible for such thoughts to gather substantial momentum.

The year 1983 will be remembered as the year that changed all that. President Ronald Reagan's March 23, 1983, speech calling for a new national effort to explore the feasibility of defense against ballistic missiles and subsequent efforts to implement this directive, which culminated in the Strategic Defense Initiative (SDI), compel strategists and politicians alike to set aside old prejudices and to take the prospect of strategic defense seriously once again. As a new debate is stimulated in the United States, it is reasonable to expect Western European leaders and strategic analysts also to return to the issue and to assess the significance of the SDI for them. They will be interested in many of the same questions raised in the United States. Can new SDI technologies offer greater prospects for protection? What might such a development mean for deterrence? Will the new U.S. initiative jeopardize the 1972 treaty regime? How might the Soviets respond? These leaders and analysts also will be interested in questions that have a particularly European slant. Can new BMD protect them against intermediate-range ballistic missile (IRBM) attacks? Does a U.S.-Soviet BMD race give the superpowers greater incentives to confine a nuclear war to Europe? How does BMD interact with the dynamics of extended deterrence?

Although it is premature to offer definitive answers to any of these questions, it is important to think them through, if only in a preliminary way. The observations in this chapter are intended to offer an initial analysis of the problems raised by BMD for alliance security. This analysis is admittedly from a U.S. perspective. It is true that the United States, as alliance leader, must be sensitive to European perspectives if it is to manage the SDI in ways that enhance alliance security and solidarity, but it is equally true that NATO allies must be willing to appreciate U.S. perspectives on the potential benefits and liabilities of moving toward a stronger reliance on ballistic missile defense. The first issue to be addressed is the relationship of the SDI to the objective of extended deterrence upon which much of NATO strategy rests. Next I will examine some of the technical requirements for SDI, with particular emphasis on defending against the threat to NATO allies. Following from this is a discussion of antitactical ballistic missiles (ATBMs) as a particular technical response to the threat to the allies. I will then address two other political aspects of the issue: the impact of SDI on arms control and the nature of the European debate about the SDI.

Extended Deterrence and BMD

NATO's strategy of flexible response cast U.S. strategic offensive weapons in a special role. Deterrence rests on NATO's capacity and will for three levels of response: direct defense, deliberate escalation, and general nuclear response. The enemy is thought to be deterred by the prospect of a credible NATO capability to defeat an aggression at the level it is initiated (direct defense); by the prospect that should direct defense fail, NATO is prepared to raise the level of violence in order to reverse the defeat (deliberate escalation); and by the prospect tha if all else fails, the United States would authorize the use of strategic nuclear forces against the Soviet homeland (general nuclear response).

In this deterrent posture, the strategic offensive forces of the alliance play a crucial role as the ultimate guarantor of the credibility of the deterrent. Indeed, it has become a tenet of Western strategic thought that any development that might erode the ability or the will of the United States to use its strategic nuclear forces in this ultimate stage of general nuclear response seriously undermines NATO's total deterrent posture. Implicit in this formulation of deterrence is the fear that if the Soviet Union comes to believe that successful aggression against Western Europe would not result in nuclear retaliation against the Soviet homeland, it will no longer face the most important disincentive NATO can offer against aggression. Implicit also is the proposition that there are circumstances in which the United States might plausibly use its nuclear forces against the Soviet Union in response to aggression against Western Europe.

Both of these assumptions have from time to time been questioned. There is at least a case to be made that an equal, if not greater disincentive to aggression would be the certainty that aggression in Europe could be made so costly through direct defense that the specific objectives of the aggressor could not be achieved within reasonable cost. Such an interpretation of flexible response would place equal weight on improving and strengthening NATO's conventional forces across the board.

The second assumption also has been called into question on occasion, although it is more difficult to assess from an objective standpoint. To those who doubt that any U.S. president would ever authorize the release of nuclear weapons if the United States were not under direct nuclear attack, the rejoinder has usually been of two kinds. First, U.S. interests in Western Europe and the vulnerability of frontline U.S. ground forces to direct attack at the outset of a European conflict guarantee that any East-West conflict would immediately engage vital U.S. interests, which might well be defended with nuclear weapons. Second, what deters the Soviet Union is not the certainty that the United States would attack the Soviet Union with nuclear weapons if NATO's collapse were imminent, but the possibility, no matter how slight, that the United States might do so. That is, deterrence is based not on certainty but on uncertainty. As long as the consequences of miscalculating U.S. actions are so catastrophic as to negate any possible assured benefit to be had from aggression, even small uncertainties can dominate policy deliberations, and deterrence remains intact.

But even those who raise such questions about the viability of flexible response would not argue that U.S. central-strategic systems could be eliminated instantly without profound consequences for deterrence. For this reason alone it is obvious that strategic defense—which in effect reduces or denies the ability of an attacker to use ballistic missiles against a defender—can have important implications for extended deterrence.

But just what are these implications? Observers come to different conclusions. Some argue that the SDI would strengthen extended deterrence because the stronger the shield behind which the United States sits, the more willing it would be, and appear to be, to take measures that might provoke Soviet strategic retaliation on the U.S. homeland. Thus, for instance, in response to a Soviet conventional breakthrough in Western Europe that seriously threatened the collapse of the NATO defense, the United States might be more willing to threaten or execute a strategic attack on the centers of warmaking power deep within Soviet territory. Such an attack would certainly provoke a retaliatory blow on the U.S. homeland, but even a relatively ineffective SDI system could significantly reduce the damage to U.S. strategic assets or society resulting from this retaliation. In this way, it is argued, the credibility of U.S. strategic retaliation could be enhanced, as would deterrence.

There is a second possible conclusion. Some observers argue that behind an SDI shield the United States would become increasingly reluctant to take those risks necessary to extend deterrence to its allies. This line of reasoning suggests that the deployment of an effective BMD shield would allow the United Stats to gradually withdraw from its overseas defense commitments, secure in the knowledge that is ultimate security could not be threatened.

A third possible conclusion assumes continued U.S. commitment to European allies, but suggests that an SDI race between the superpowers would eventually lead to a situation where extended deterrence would be undermined by an increased willingness of both superpowers to engage directly in conflict in the theater, knowing that each resides behind an impenetrable shield.

Each of these conclusions requires careful examination. The first argument—that BMD can only enhance extended deterrence—rests on a number of plausible but uncertain assumptions. For example, it presumes the enhanced credibility of the threat of strategic retaliation in response to theater aggression, but it ignores the implicit possibility that the dynamics that enhance the credibility of such retaliation also might make provocations and aggression in the theater less costly for the United States, and hence may actually weaken deterrence. If, when contemplating a course of action in a European crisis, U.S. leaders foreseen the eventual possibility of Soviet attack on the U.S. homeland, they may view the possibility with less alarm from the vantage point provided by an effective SDI. What would make for a stronger deterrent against Soviet aggression may serve to expand U.S. options in the theater, hence *lowering* the deterrent threshold in Washington.

The second conclusion seems to imply that the only reason the United States has defense commitments overseas is because its interests overseas are such that an attack on them may eventually leave the United States vulnerable to attack. But it is clear that important *positive* U.S. interests are engaged in Europe and Asia—not to mention elsewhere—even were the United States to be completely invulnerable to attack. Indeed, until the late 1940s the United States, for all practical purposes, was invulnerable to attack and yet found itself pulled inexorably into involvement in the two major world wars. U.S. political and commercial interests overseas are such that much of the U.S. population is happy to protect them from attack, irrespective of the vulnerability of the U.S. homeland to attack by Soviet ICBMs.

Behind the third conclusion rests an assumption that actually underpins the previous two as well—that a BMD system could be developed to render each superpower virtually invulnerable to a nuclear attack. But even the most optimistic technical assumptions about the effectiveness of SDI against ICBMs and submarine-launched ballistic missiles (SLBMs) would not eliminate the threat to each superpower inherent in its nonballistic means of weapon delivery. Aircraft and cruise missiles—not to mention "suitcase bombs" and other esoteric means of delivery—could be effective penetrators of any SDI shield, and hence the SDI will not in itself relieve either superpower from a requirement for strategic offensive deterrent forces.[2] Indeed, a serious SDI race may stimulate a move to increased reliance on air-breathing systems for purposes of retaliation.

The upshot of all this, perhaps not surprisingly, is that the effect of SDI on extended deterrence is not clear-cut or easy to foresee. Much depends on the effectiveness of an SDI system, what it protects, what threat if confronts, Soviet capabilities, the implications of SDI for the independent European deterrent powers, other elements of the East-West balance, and the political context of East-West relations.

Perhaps the most difficult aspect of the SDI for European governments to accept, however, is the long-term vision of the president—that the SDI can be part of a process to eliminate nuclear weapons from the face of the earth. For in spite of the fact that the SDI is unlikely to provide a perfect defense against ballistic missiles—not to mention air-breathing systems—political rhetoric that places the SDI in the context of overthrowing the deterrence relationship upon which stability is currently based poses formidable challenges to European political elites. These elites long have accepted the need for an offensive nuclear guarantee by the United States to offset the conventional superiority of the Warsaw Pact. Only recently and at considerable political cost were these governments forced to develop a public consensus within their countries regarding the important role offensive nuclear weapons play in NATO's deterrent, as part of the implementation of the INF decision. Rhetoric that challenges these long-standing beliefs not only undermines the public consensus emerging from the INF decision; it goes against the strategic instincts of most European political leaders and strategic analysts.

It is in part for this reason that the Reagan administration has deliberately toned down the rhetoric of eliminating offensive deterrence that characterized the early U.S. discussion of the SDI and has replaced this rhetoric with themes that fit more comfortably into traditional notions of deterrence and stability.[3] SDI now is described more as a way of enhancing deterrence by rendering offensive first-strike scenarios less certain of military success. This theme is generally more attractive to political leaders in Western Europe and gives rise to interest in more limited, traditional notions of target protection, particularly the ATBM. Before turning to the ATBM problem, however, it is important to understand the basic technical functions of the SDI.

BMD Effectiveness

Any SDI system must be able to destroy ICBMs or their warheads (reentry vehicles, or RVs) at some point between launch and impact on target. As such it must be capable of five related functions: target acquisition, target tracking, discrimination, interception, and target kill.[4]

Target acquisition determines that an ICBM has been fired and knows where it is. It is usually accomplished through sensor systems that detect an enemy missile launch and relay the message to the National Command Authority (NCA). Target tracking is a somewhat different function involving the capability of a sensor to follow the target through its flight path and on the basis of its trajectory to predict where it will be at any moment during its flight. This involves a complex integration of sensing and computer technologies, which calculate highly complex equations of motion. Discrimination is another essential function. The offense can be expected to include a number of decoys or other physical objects to confuse the defense into attacking the wrong objects, thereby wasting precious time and interceptors. It is complicated by physical laws; in the vacuum of space, objects of different weight and shape will fall along the same trajectory at exactly the same speed because there is no friction in space to work differentially against these objects. Target interception involves the capability to use all the information acquired about the potential targets to launch an interceptor, or some other sort of kill mechanism, so that it will cross the path of the target at close enough range so that the kill mechanism will destroy or render useless the target. Finally, target kill involves the successful initiation of a mechanism—a nuclear explosion, a conventional explosion, or some other, more exotic mechanism, such as a particle or laser beam—that will cause the target to be destroyed or otherwise rendered harmless.

The five complex functions are performed against the targets with very distinctive flight characteristics. An ICBM and its RVs pass through four phases of flight. The first phase is the powered stage of flight, lasting from 30 seconds to several minutes, called boost phase. The power for the total flight is imparted during this period; during the rest of the flight, the missile and its RVs are in a free fall back to earth, arcing in a ballistic trajectory. The second phase is sometimes called postboost; it always involves the separation of the RV from the missile, but it may also involve, with a MIRVed system, the separation of a MIRV-dispenser (known as a postboost vehicle, or PBV) from the ICBM and various maneuvers of the PBV as it releases its RVs along the appropriate trajectories. It also involves dispensing decoys and various other objects to spoof any defense or early warning system. This phase may also last several minutes.

Next, the RVs move into midcourse—a relatively long period in which they coast along their individual flight paths in free fall, accompanied by any decoys or chaff the offense has chosen to use. It lasts some 15–20 minutes.

Then the reentry vehicles move into reentry. For the RV, this is the most stressful period of flight. Reentering the atmosphere at 17,000 miles per hour is often likened to smashing into a brick wall. The RV must be designed to withstand tremendous physical forces—air resistance and friction—with two major objectives: surviving the tremendous heat created by this friction and falling through the atmosphere at a relatively predictable trajectory so that accuracy can be maintained. During this phase, which lasts only a few minutes, the RV will decelerate enormously; so will any decoys, but usually at a faster rate, which facilitates the function of discrimination.

Traditional BMD systems have focused on intercept in the reentry phase or late in the midcourse phase, with ground-based, nuclear-typical interceptor missiles deployed fairly near the targets being defended. The traditional approach is quite stressful. It places a heavy premium on high-acceleration missiles and on very accurate interception and reliable target kill. Once the RV has moved into reentry phase, there is precious little time to compensate for any errors. But a system that could overcome these hurdles would have numerous advantages. The systems could be located in such a way as to defend higher value targets preferentially and also use the atmosphere to help perform the discrimination function. The whole system can be located on the ground and need not concern itself with operations relatively far away from the objects being defended.

Another way has been proposed to deal with the discrimination problem in such a way as to handle the problem of system saturation: boost-phase intercept. The concept is based on the principle that if a system can destroy ICBMs while they are in boost phase (i.e., before RV separation), there is no need to worry about discriminating RVs from decoys and chaff; all will be destroyed in one interception. Furthermore, boost-phase defense reduces the value of MIRVing as a means for enhancing penetration because one interception destroys all warheads aboard a booster. Indeed, an effective boost-phase system would provide incentives for de-MIRVing, which many judge to be a stabilizing development.[5]

The problems of boost-phase interception are daunting. Although target detection and tracking may be easier because of the highly visible signature of booster burn, boost-phase intercept requires placing sensing and computational assets in a location to see the boost phase virtually from the moment of launch. This requires at least some space-based assets over Soviet territory at the moment of launch—an expensive and technically demanding proposition. There are two ways in theory to do this. One would be to station sensing and kill assets in space over the Soviet Union at all times. This would require numerous satellites, far more than the number required simply to cover Soviet missile fields, because as the satellites orbit so much of the satellite population would be over non-Soviet territory for so long. Geosynchronous satellites would solve the number problem because they would not wander away from Soviet territory, but they could be too high to perform their tracking and kill functions effectively. Boost-phase defense also would require space assets that would be relatively invulnerable to attack themselves. Otherwise, they would be such important targets for enemy preemption that one could imagine any conflict beginning with the destruction of these assets. Defensive measures to protect space assets are extremely difficult to devise. Some argue that a space-based, boost-phase system could defend itself by its very nature. But there are significant differences between BMD and counter-antisatellite (ASAT) weapons that cannot be ignored.

The other approach would be to wait for a crisis—or even warning of ICBM launches—before lofting a boost-phase system into position. This approach would reduce boost-phase vulnerability, but it puts tremendous strain on the warning and launch system because of the very short times involved.

Boost-phase interception, although extremely difficult, is of interest for several reasons. First, it is one of the new technological concepts the Reagan administration is exploring. For that reason alone it would be worth some mention. However, there is another point here. Boost-phase systems make no obvious distinction between RVs aimed at the United States or RVs aimed at Europe. For this reason, it is often said that a boost-phase system successfully deployed would protect both the United States and its allies from attack. As such, it would obviously enhance extended deterrence.

Such a conclusion rests on the assumption that ballistic threat to European allies comes increasingly from mobile systems. The SS-20 is an obvious example. Target acquisition in boost-phase intercept is much easier for fixed ICBMs than for mobile ICBMs. Against fixed ICBMs, one knows beforehand where to point relevant sensors. Mobile systems, on the other hand, may be anywhere within a particular area; thus, sensors must look into any event that may look like a missile launch. Now this is certainly possible, but it is more difficult, more time consuming, and more prone to error than target acquisition against fixed ICBMs.

There are other potential problems even a boost-phase system would encounter in protecting European targets. The flight times of ballistic missiles from Soviet territory to European targets are much shorter than they are for intercontinental trajectories. This places increased time strains on a boost-phase system and may in effect rule out the launch-in-crisis approach for a space-based defensive system. Furthermore, many of the ballistic missiles facing European allies are of much shorter range than the SS-20. Some have ranges on the order of 310 miles and thus must be deployed forward in Eastern Europe before they can be used against targets in NATO Europe. This not only increases the time constraints on a boost-phase system defending Western Europe; it also increases the relevant area to be scanned by a sensor for target acquisition and probably requires more satellites as a consequence.

All this is not to say that there is not an element of validity in the assertion that boost-phase SDI need not distinguish between the threat to the United States and the threat to its allies. This is certainly true in a first order sense. If some SS-18s or SS-19s were allocated to European targets, then a boost-phase defense against these missiles would make no distinctions. However, the ballistic threat to Europe is of a broader, more complex variety than that to the United States, and an SDI system designed to provide boost-phase protection for both must take these differences into account.

There is an alternative approach to defending European targets that is receiving renewed attention because it relies on relatively traditional technologies and is not beset by the uncertainties facing the boost-phase approach. Research into antitactical ballistic missile concepts also is not explicitly prohibited by the 1972 ABM Treaty. One can imagine traditional ground-based sensors and interceptors deployed in Europe to destroy IRBM or even shorter-ranged weapons in reentry. They would perform all the same functions as do traditional BMD systems and would use the same technologies. However, there would be one major difference deriving from the technical characteristics of the attacking RVs. Shorter-range RVs tend to have much shorter flight times. This alone would stress the acquisition and tracking functions significantly. Short-range RVs also spend less time in space and more time in the atmosphere, thereby causing their flight paths to be less predictable. All these problems suggest that an ATBM would find the most useful mission in the protection of specific military "point" targets, rather than area or population targets, because most interceptions would take place at such low altitudes that soft targets might be vulnerable to the heat, blast, and radiation from the interception.

This admittedly brief, oversimplified review of SDI technologies suggests general propositions regarding the ability to protect Europe.

- Boost-phase or postboost-phase systems can provide Europe with the same degree of protection against ICBMs as they provide the United States. This is because in boost or postboost phase the ultimate destinations of warheads are not known; it is just as easy to go after all potential RVS before they separate as it is to wait until it is possible to determine their individual destination targets.

- Midcourse or reentry-phase defense systems require a more explicit determination to intercept RVs heading toward European targets if Europe is to be protected. For reentry-phase systems, such a decision would probably require SDI deployments on the European continent.
- Taking into account timing and trajectory phenomena, midcourse and reentry-phase protection of European targets imposes highly stressful technical demands on any proposed system.

Just how effective would an SDI system, such as an ATBM, have to be in order to offer an interesting protection to European targets? The answer depends in part on what the United States would choose to protect. One possible target set for protection is European urban areas. Indeed, this is the most compelling target set for a number of reasons. First, most people would expect that an expensive SDI system would be able to offer individuals some degree of protection. Moreover, European territory is heavily urbanized—thus, many important military targets are located in or near major population centers. Protection of these targets in effect may require some form of urban protection.

It is worth considering the military target set in Europe in more detail. For example, there are 200 airfields in NATO Europe, more than twenty nuclear storage sites, and more than thirty major divisional and corps headquarters. Each of these would be an important military target for the Soviet Union in the early days of a conflict. Indeed, we know from Soviet military doctrine and exercise data that airfields, nuclear storage sites, and port facilities are among the top priority targets for attack in the very early hours of a war. Of the 300 major NATO targets in Europe, perhaps 100 of them are of highest priority and would merit consideration as active ATBM defense sites.

Would an ATBM system offer significant protection against an attack on these targets? Let us specify the most likely technologies for an ATBM—a ground-based traditional endoatmospheric/intercept system, with some mobility in radars and launcher/interceptor vehicles. Mobility would serve two useful purposes: It would enable an ATBM unit to follow the target it is protecting around if it were mobile (e.g., Pershing II or ground-launched cruise missiles—GLCM), it would also make the offense's efforts to locate and suppress defenses more difficult.

Such a system, if technically reliable, could probably force the enemy to allocate more than two RVs per target defended; the enemy would probably not be able to achieve an attack price of more than eight RVs per target. A more conservative estimate would be three or four RVs per target.[6] This would result in the Soviets allocating an extra 100–200

RVs to a nominal critical, defended European target set. Such numbers, of course, are not difficult for the Soviets to generate.

ATBM units could always be deployed on a larger scale to protect all 300 or so of NATO's targets of greatest interest to Soviet military planners. Whether or not it would make sense to do so would depend largely on the cost of defenses.

The utility of ATBMs cannot only be judged by this calculus. One also must consider the nature of the nuclear threat facing NATO Europe. To a far greater extent than the United States, NATO Europe faces a major "air-breathing" threat from bombers and fighter bombers based in the Soviet Union and Eastern Europe. Furthermore, Soviet forces will probably be receiving a new cruise missile arsenal within the next five years.

These slow-flying systems generally pose less of a threat to time-urgent targets, given the relatively long warning period NATO would have prior to attack. Air assets at forward airbases could be moved back, and Pershing and GLCM launchers could be moved off base to relatively more protected areas. But much of the NATO target base is not time urgent. Major wartime command posts, nuclear storage facilities, command, control, communications, and intelligence (C^3I) modes, and port facilities are all vulnerable to the low-flying aircraft like Fencer, Flogger, or Fitter, not to mention the heavier Badger and Blinder bombers and the more modern Backfire. As a general proposition, however, an ATBM could be ineffective against aircraft and cruise missiles because of the difficulty of predicting flight profiles, the countermeasures any modern aircraft will use to counter defenses, and the greater complexity of the environment at lower altitudes during war. Because so much of the NATO Europe target set is vulnerable to such air-breathing systems, which are numerous on the Soviet side, it may be pointless to pursue ATBM coverage for all targets across the board, unless the context of a broader effort to improve NATO air defense capabilities.

Is it more useful to consider ATBM Long-range Intermediate Nuclear Forces (LRINF) bases, attack on which could have unfortunate escalatory consequences? In some ways, LRINFs are time urgent; their mobility was introduced so that in a crisis they could be moved off base to hidden locations, thereby reducing the opportunities for preemption. Thus, if the system worked according to specification, there would be no need to deploy ATBMs around LRINF bases; the missiles would be long gone before the first attacking RVs hit.

But, argue advocates of ATBMs for LRINF, political leaders may be reluctant to authorize dispersal of LRINF missiles off their bases because such an act would betray to the Soviets such an advanced state of readiness that it actually might provoke hostilities. Furthermore, even if war broke out before LRINF missiles were dispersed, subsequent dispersal might be delayed because of the fear that it would signal the intention to use nuclear weapons prematurely. Indeed, there may be numerous reasons NATO leaders would want to postpone the decision to disperse for as long as possible. An ATBM system deployed to protect Pershing and GLCM bases, it is argued, would give NATO leaders valuable additional time to deliberate prior to dispersal or launch.

Let us examine this argument in greater detail. The relevant target set to be protected is relatively well known. Some 464 GLCMs will be deployed among six bases in five countries—the United Kingdom, West Germany, Italy, Belgium, and the Netherlands. In addition, three bases in West Germany will deploy a total of 108 Pershing II missiles. All the bases are well known to Soviet intelligence; none is likely to be more than 25 square miles in area. One 300 kiloton (kt) weapon would cover such an area with at least 5 pounds per square inch (PSI) maximum overpressure, which should be enough to destroy virtually all missiles at the base. (These LRINF missiles are extraordinarily soft and vulnerable to blast effects.) Two warheads would be enough to give confidence to even a conservative Soviet military planner. Note that the optimal strategy for the Soviet planner is not to target each NATO missile separately; the weapons are sufficiently soft and close together to make a more general strategy far more attractive.

In an undispersed mode, the most that an ATBM would be able to do is to force the enemy to use more RVs than it would normally do to destroy NATO LRINF facilities. Given the extreme vulnerability of these unprotected missiles prior to dispersal and the small number of bases, the ATBM must provide strong leverage indeed before it becomes attractive. An ATBM defense that forces the Soviets to use ten RVs per base in order to have high confidence of destroying it increases requirements for a Soviet planner, but the trade-off ratio is still attractive: For 90 RVs spent—some 30 SS-20s—the Soviets can destroy 572 U.S. warheads. Forcing the ratio as low as possible may be one NATO goal. At the extreme, the Soviets may want to associate one ATBM unit per missile, a unit being defined as an interceptor and an independent radar unit for target acquisition, tracking, and battle management. Such an approach would require 572 ATBMs. It would be very expensive, and it may be vulnerable to exotic offensive tactics. But it might force the Soviets to use at least 590 RVs, or 162 SS-20s, to attack and destroy the NATO LRINF force.

So far I have been discussing the use of ATBMs to protect against attack by nuclear ballistic missiles. But many Soviet short-range ballistic missiles (SRBMs) that threaten Europe today have a conventional capability as well and could be used against point targets in an initial conventional phase of a Soviet attack. Many of these NATO targets vulnerable to conventional attack of this sort are nuclear related—for example, nuclear storage sites and INF bases. Such vulnerability to conventional attack could place tremendous escalatory pressures on NATO leaders, which would be well worth minimizing if possible.

Does an ATBM concept make more sense against conventional ballistic missile attack than against nuclear attack? No clear-cut answers can be provided in the absence of a system with specified capabilities and costs, but some preliminary observations can be ventured. First, the attacker's requirements for successful conventional missile attack are different and in many ways more stressing than for nuclear attack. Conventional munitions require delivery directly on target for success because their destructive effects tend to be highly localized. For the same reason, one conventional warhead can at best destroy only one target. Against an airbase, for example, two nuclear RVs may be enough to put sufficient PSI, blast, and heat to an area of 25 square miles to assure the destruction of most targets in the area. In contrast, several conventional munitions may be required to put a single runway out of commission; more would be required for additional runways at the airbase, not to mention aircraft hangers and control buildings. The total number of conventional RVs required to put a NATO airbase out of action could range from five to ten and perhaps more. Each of these RVs would have to be directed at specific targets on the airbase, and they would have to be of sufficient accuracy to compensate for the localized destructive effects of the munitions. Against a nuclear storage bunker, which is already hardened against conventional munitions, several direct hits may be required to assure destruction; the same is probably true for the hardened shelters housing GLCM launchers on base.

This means that a comprehensive conventional ballistic missile attack against NATO airfields, nuclear storage sites, and INF bases might require between 1,200 and 2,000 conventional warheads, which is well beyond current Soviet SRBM capabilities. A selective attack on fifty of the most important NATO airbases, along with nuclear storage sites and INF bases, might require 400–600 such warheads. This kind of target set is, of course, well within the capabilities of the 800–1000 SS-21, SS-22, and SS-23 missiles deployed in the Soviet Union and Eastern Europe. It is, therefore, against this type of attack that any consideration of ATBM protection should take place.

The defender's objective might reasonably be to force such expenditure of warheads that it places an unacceptable strain on Soviet SRBM capabilities. From this perspective, even doubling the number of RVs required to destroy a protected target may suffice to make the attack unattractive. Ground-based interceptor missiles, tied to ground-based and air-based sensor networks, would be one configuration worth considering. The scope of the ATBM network would then be dependent in part on the "kill probability" (P_k) of each interceptor missile. A P_k of .7 would require 850–900 interceptor missiles to destroy the first 600 incoming conventional RVs; the second 600 would essentially have a "free ride," but the effect of the defense would have been to double the required level of attack, thus achieving NATO's objective.

Is such a P_k possible? The answer depends on several factors.

- Timely warning of attack and sufficient data processing to permit engagement of the income RV before it hits its target.
- An effective kill mechanism. Given the short flight times and trajectories of SRBMs, it is unlikely that nuclear-armed interceptors could effect interception without unacceptable damage to the target being defended. Conventionally armed interceptors would seem to offer the best solution. These would have to be accurate enough for the localized conventional effects to destroy the warhead or send it sufficiently off course to prevent damage to the target.
- Invulnerability of sensor and data processing nodes. Precursor attacks capable of destroying these nodes could render the interceptor missiles themselves useless or seriously degrade their P_k.
- Resistance to inexpensive offensive countermeasures, such as electronic countermeasures (ECM) or chaff, which would also have the effect of reducing P_k.

Whether these conditions can be met remains uncertain, but the issue is worth further study. If a robust, effective force of 850–900 ATBMs could be fielded to offer protection against an attack on NATO's most important airfields, nuclear storage sites, and INF bases, this could in effect deny the conventional ballistic missile option that currently concerns NATO planners.

But knowing that an ATBM system can deny the conventional attack option against an important array of NATO nuclear targets does not imply that ATBMs provide the best way of doing so. In practice, any workable ATBM system will have to be compared to other available options, such as dispersal, hardening, revamping of the alert machinery, and the various other methods to defend against these types of attacks. The best solution is unlikely to be a "pure" one; rather, it will incorporate a variety of reinforcing partial solutions.

To summarize this discussion:

- SDI concepts can be extended to provide some degree of protection to the European allies, at least in theory.

- The earlier the interception takes place, the more irrelevant is any distinction between the threat to the United States and the threat to its allies. A boost-phase intercept system does its work before it knows where RVs are heading.
- For the protection of European society, the relevant SDI concepts would be so expensive as to be prohibitive, even if they could be made to work effectively.
- Protection of military targets in Europe may be a more feasible objective, particularly if the threat is from conventional ballistic missiles; but an effective ATBM defense against such a threat must be compared to other options for denying the enemy this capability.

Other Considerations

Technology, however important, is never the only element in Western defense debates. The question "Will it work?" even if adequately specified and qualified, can only be part of the answer to a public concerned not only with matters of technical feasibility but also with matters of political and strategic wisdom.

By what criteria are U.S. allies likely to assess the political and strategic wisdom of ballistic missile defense? One important criterion, perhaps more political than strategic, will be the impact of such a program on overall East-West relations. European governments are responsive to general public concerns about the state of East-West relations, particularly U.S.-Soviet relations. In an atmosphere of East-West cooperation, it is conceivable that the superpowers could gradually come to terms with the strategic impact of new technologies and cooperate in making adjustments to their strategic force postures in order to promote stability during a transitional period. This could entail renewed momentum in the Strategic Arms Reduction Talks (START), which aim to bring offensive forces under strict controls so as to suppress the natural tendency a superpower might have to offset strategic defenses by increasing offensive force levels.

Whether this will be a result of the improved U.S.-Soviet relations emerging from the November 1985 Reagan-Gorbachev summit remains to be seen. Although relations between the two sides have improved and negotiations in Geneva seem to be moving ahead in strategic offensive reductions and INF, the two sides seem at an utter impasse regarding the SDI. The U.S. position is that research in this area is intrinsically important and within the constraints of the ABM Treaty; the Soviet position is that any research falls afoul of the ABM Treaty, is strategically destabilizing, and will only stimulate offensive arms increases. By permitting detailed work in the other areas of the negotiation to continue while there is such a fundamental gap regarding the SDI, the Soviets have chosen not to pursue the sort of crude SDI linkage with which it would be relatively easy for the West to deal. Yet they have not foresworn a more sophisticated, drawn-out kind of linkage that develops tantalizing prospects of eliminating nuclear weapons before the turn of the century while at the same time holding open the option of throwing the SDI back into the works at a tactically opportune moment.

It is this more subtle, more damaging political strategy that the Soviets could bring to bear on the United States and its allies regarding the SDI in the coming months. Here the potential for division is high, unless the United States can make a credible case that Soviet intransigence, not U.S. resistance, is blocking the prospects for an agreement in the interests of both parties.

In addition to these general political concerns, there may be a relatively unique strategic debate. One could imagine a European debate among those who believe that strategic defense undermines stability and deterrence and those who believe that it supports and enhances stability and deterrence. Critics of SDI will point to the inherent capability of an SDI system to limit retaliatory damage that the other side could inflict, thus lowering the threshold of deterrence. They also will point to the role an imperfect SDI system could play in supporting a first-strike offensive mission, thereby effectively blunting a "ragged retaliation." They also will note that an imperfect defense might serve only to stimulate further growth of offensive arsenals in an effort to counteract the defense.

SDI advocates—who are likely to be smaller in number in Europe than in the United States—will respond that there is nothing necessarily stabilizing about the assured destruction capability; that even limited defenses could serve useful functions simply by complicating Soviet first-strike planning; and that the Soviets already are involved in a massive program of their own, which if left unchallenged would be far more destabilizng than an SDI race. They will argue that the SDI can be made sufficiently effective to reduce the value of offensive ballistic missiles to such a low level that both sides will be willing to trade them away in arms control agreements.

In these respects, there are likely to be some strong resonances between the U.S. debate and the European debate. But there will be certain unique emphases in the European debate. For example, Europeans will be concerned with the implications for flexible response from a perspective that many in the United States may not appreciate. They will point out that NATO's deterrence has rested for more than two decades on the notion that NATO may use nuclear weapons first to defend against a conventional onslaught and that truly effective SDI programs could eliinate this option for both sides, or so reduce its value that such a threat will no longer seem to be an effective deterrent. Hence, the probability of conventional war may be seen to increase. To Europeans, the prospect of a major conventional NATO–Warsaw Pact engagement is only slightly less horrifying than the prospect of a nuclear war. Indeed, European support for the policy of nuclear deterrence has rested on the premise that it reduced the likelihood of *any* war occurring, whether nuclear or conventional. A defense posture that made conventional war more likely, even at the expense of making nuclear war less so, is not likely to receive overwhelming support from European publics or their governments.

There are several responses to this concern. One is to point out that the SDI will probably not totally eliminate the prospect of significant offensive nuclear attacks, either in retaliation or as preemptive first strikes. This is likely to be a true statement whatever the outcome of U.S.-Soviet defensive competition. SDI will not address the air-breathing threat, nor is it likely to work as well against short-range ballistic missiles and SLBMs as it would against ICBMs or IRBMs. Indeed, the development of a sophisiticated SDI by both sides may redirect offensive force competition. This point, however, runs the danger of undercutting one of the principal rationales for the U.S. program—to shift away from deterrence based on retaliation to stability based on defense. The most effective response to this concern is to point out that nuclear weapons have always played a relatively limited role in the deterrence posture of the alliance; that conventional forces have always played an important role in conventional deterrence; and that improvements in NATO's conventional forces can more than offset any diminution of conventional deterrence that results from a relatively successful BMD program.

Another specifically European strategic concern will be felt most strongly in Britain and France, Western Europe's only nuclear powers and the only U.S. nuclear allies. Both countries have maintained small, expensive national nuclear forces for almost three decades. Each places enormous political weight and prestige on the mission performed by these forces. Neither would be agreeable to developments that cast into doubt the ability of these forces to penetrate with confidence to their intended targets.

Both countries recently faced difficult decisions regarding the modernization of their strategic deterrents. In the process of this decision-making the issue of penetration of the new forces has been a central consideration. It is reported that this was the main reason for Britain's choice of the Trident over cruise, and it certainly was the motivating factor in the decision to develop, at a reported expense of $1 billion, a new warhead to penetrate Soviet defensive deployments (Chevaline). French decisionmaking has been less open, but the choice of the rather expensive SLBM program in the future must reflect these same concerns to some degree, especially because less-reliable alternatives are much less expensive.

Given this background, it should come as no surprise if the governments of Britain and France harbor deep concerns about U.S. and Soviet defensive programs that none of the other allies shares. Specifically, if a new U.S. BMD effort stimulates a major U.S.-Soviet defensive race—or even if the superpowers agree to cooperate in mutual, limited deployments—the degree of protection afforded by these systems may not be perfect with respect to large superpower arsenals, but may be rather daunting to the much smaller British and French arsenals.

What level of Soviet BMD protection would begin to severely constrain the British or French nuclear forces? Would these two countries be able to respond appropriately within financial constraints? These questions are well beyond the scope of this chapter, but several points are worth considering.

- First, to the extent that these developments pose real problems for Britain and France, they should be concerned irrespective of U.S. activities. Soviet R&D activities present an ongoing problem for British and French force planners and would continue to do so even if all U.S. programs were halted immediately.
- Second, perhaps more importantly, there may be a tendency for the two countries to exaggerate the problems posed by Soviet strategic defenses. Some have suggested that with any reasonable criteria for target coverage, Britain and France are likely to be able to meet these criteria whatever the Soviets deploy.[7]
- Third, the United States should be particularly careful to avoid the impression that its own activities could make matters appreciably worse for its allies.

Given this array of concerns, it is probably unrealistic for the U.S. public to expect initial enthusiasm on the subject of BMD from European allies or publics. The overwhelming sentiment has been and will continue to be one of concern. Perhaps the best that can be expected is that this concern will not be transformed into more overt hostility regarding the project.

Conclusions

Assuming that European enthusiasm for the SDI will inevitably be somewhat weaker than U.S. enthusiasm, are there any steps that can be taken on both sides of the Atlantic to minimize the opportunities for discord on this subject? Several come readily to mind.

1. The United States must be cautious not to oversell the SDI. It should be explained for what it really is—an advanced research program to validate technical concepts, not a grand effort to rid the world of offensive nuclear weapons. The latter approach runs into two serious dangers. It may reflect technical optimism out of all measure to realistic assessments of the technical possibilities. It also touches raw nerves in doctrinal matters, where the articulation of deterrence theory has taken on something of a theological quality. If and when research validates certain technical concepts, efforts to approach the strategic and doctrinal matters will take on a far richer meaning and will be more fruitful. A low-key undersell is perhaps the best way to avoid alarming European publics and governments, to avoid premature conflicts about doctrine, and to give the U.S. R&D community time to examine all the relevant technologies without excessive glare from a political spotlight.
2. A second step is for the United States to maintain a regular, low-key dialogue with its allies on the status of the SDI R&D program as it progresses. As we have seen, BMD raises issues of direct interest to the allies; they have a right to know as much as the United States can tell them about this program as it develops. This process in fact was accelerated at the March 1985 Nuclear Planning Group (NPG) meeting, where Secretary of Defense Caspar Weinberger invited colleagues to join in the program. Intensive bilateral discussions to date have resulted in a formal U.S.-UK agreement on cooperative research. Other similar agreements are likely to follow. This provides a channel for European information about the program and enables European industry to take advantage of the new technological efforts in this area.
3. At the same time, an important third step is for the European allies to approach this dialogue with an open mind; they should avoid engaging the doctrinal and strategic issues before it would be useful to do so.
4. Finally, a requirement for harmonious handling of this issue within the alliance context would seem to be sustained improvement in U.S.-Soviet relations—or if this proves impossible, a credible U.S. record on behalf of such an effort. Nothing raises anxiety levels throughout Europe quite so much as a U.S.-Soviet chill. In a chill, almost any initiative is viewed as provocative, risky, and

dangerous. In a thaw, initiatives stand a better chance of being judged, in the West at least, on their own merits.

Notes

This research was funded in part from a NATO fellowship. The views expressed herein are solely those of the author and should not be attributed to the State Department, the International Institute for Strategic Studies, or any other institution or individual.

1. Ashton B. Carter and David N. Schwartz, eds., *Ballistic Missile Defense* (Washington, D.C.: The Brookings Institution, 1984), p. 344.
2. Speech by Paul Nitze before Philadelphia World Affairs Council, February 20, 1985.
3. See, for example, U.S. Department of State, *The Strategic Defense Initiative,* Special Report 129 (Washington, D.C.: U.S. Department of State, Bureau of Public Affairs, June 1985).
4. A good summary, from which much of the following is drawn, is in Steven Weiner, "Systems and Technology," in Carter and Schwartz, *Ballistic Missile Defense,* ch. 3.
5. The Scowcroft Panel Report, 1983.
6. This judgment would follow from the analysis in Ashton B. Carter, "BMD Applications: Performance and Limitations," in Carter and Schwartz, *Ballistic Missile Defense,* ch. 4.
7. Lawrence Freedman makes this case in "The Small Nuclear Powers," in Carter and Schwartz, ibid., ch. 7.

7

British and French Strategic Forces: Response Options to Soviet Ballistic Missile Defense

James C. Wendt

The attitude of Europeans[1] toward the U.S. guarantee to use nuclear weapons in defense of the alliance has varied during the years. At times, Europeans have feared that the United States seemed likely to use nuclear weapons too soon in any conflict in Europe, thereby leading to the destruction of Europe. At other times, especially since flexible response has been adopted as official NATO doctrine, Europeans have had doubts about whether the United States would ever use nuclear weapons to defend them.

Perhaps most Europeans have at some time had doubts about the U.S. nuclear guarantee. Two countries, however, have had the will and the capability to act on their doubts. Britain and France have developed independent nuclear forces that represent a potential retaliatory threat to the Soviet Union. Under the circumstances of an impending Soviet conquest of Europe and a U.S. failure to act, either of these countries could unleash a potentially serious nuclear attack against the Soviet Union. During the next decade or so, both the British and the French will expand and modernize their strategic forces. During this same period, however, the Soviets might deploy some form of ballistic missile defense (BMD) that would reduce the effectiveness of these attacking forces.

If the Soviets succeeded in deploying a BMD that could substantially reduce the effectiveness even of the expanded British and French forces, what would be the consequences? Would the British or the French get out of the nuclear business altogether and eliminate their strategic forces? Would either or both continue to operate a nuclear fleet that had little credibility because of its perceived potential ineffectiveness? Or could they pursue response options that would offer the prospect of increasing the potential effectiveness of their strategic forces at a reasonable cost?

British Strategic Forces

Rationale

With the purchase of Polaris missiles from the United States in 1962, the British government needed a formal rationale to justify the acquisition of this weapon system.[2] Because Britain depended on the United States for the sale of these missiles as well as for continuing aid with warhead design, submarine design, test facilities, and spare parts, Britain had to produce a rationale consistent with U.S. policy. Nevertheless, the British had not forgotten the sting of Suez and their declining place in the world, which prompted a desire for independence. Ultimately, they invented a rationale called the "second nuclear decision center" within NATO. This rationale allowed them to stress both their role in NATO deterrence—a position consistent with U.S. policy—and their independence—the unique position of British-controlled forces in deterrence.

The British nuclear force is officially committed to NATO. According to then prime minister Harold Macmillan, the 1962 Polaris agreement provides that the use of this force is "for the purpose of international defense of the Western Alliance in all circumstances."[3] Independent British decisionmaking within NATO in the use of the force is to lead to greater deterrence because

> the nuclear decision, whether as a matter of retaliatory response or in another circumstance, would, of course, be no less agonizing for the United Kingdom than for the United States. But it would be a decision of a separate and independent Power, and a Power whose survival in freedom might be more directly and closely threatened by aggression in Europe than that of the United States. This is where the fact of having to face two decision-makers instead of one is of much significance.
>
> Soviet leaders would have to assess that there was a greater chance of one of them using its nuclear capability than if there were a single decision-maker across the Atlantic. The risk to the Soviet Union would be inescapably higher and less calculable. This is just another way of saying that the deterrence of the Alliance as a whole would be stronger, the more credible, and therefore the more effective.[4]

This rationale masterfully identifies the Soviets, not the British, as the party with doubts about the U.S. nuclear guarantee. Moreover, the use of British strategic forces may be more credible in Soviet eyes because the British are closer to the threat in Europe.

No British government has ever officially expressed doubt about the U.S. nuclear guarantee. However, in the unlikely event that the United States and the United Kingdom do not define British "supreme national interests" in the same way, Macmillan qualified the phrase, "for the purpose of international defense of the Western Alliance in all circumstances" by adding "except where Her Majesty's Government may decide that supreme national interests are at stake."[5] The British government considers it prudent, under these unlikely circumstances, to prevent blackmail and deter attack by having an independent strategic nuclear force capable of inflicting unacceptable damage on the Soviet Union.[6]

British Strategic Posture

Eight years after the 1962 Nassau decision about Polaris, the British had four nuclear submarines commissioned and available for patrol—*Resolution, Renown, Repulse,* and *Revenge.* Each submarine carries sixteen Polaris A-3 missiles with a range of 2,900 miles. Each missile has three reentry vehicles (RVs) with a yield of 200 kiloton apiece.[7]

In July 1980 the British decided to purchase Trident I (C-4) missiles from the United States as a follow-on system. When the United States advanced the timing of the Trident II (D-5) program, the British changed their decision and in March 1982 agreed to purchase D-5 missiles to maintain commonality with the United States.[8]

The new Trident submarine-missile system will replace the Polaris system. Four new ballistic missile submarines (SSBNs) will be built of the Ohio-class—that is, similar in size and configuration to the U.S. Ohio-class submarines designed to accommodate the D-5 missiles. The British will design these SSBNs and will fit sixteen tubes to each submarine.[9] The D-5 missile is estimated to have a range of 6,200 miles. The design of the front end of the missile has not been announced, but it is expected to contain eight RVs.[10]

Thus, after modernization, the minimum British threat with one boat on patrol will be about 128 RVs (i.e., sixteen missiles with eight RVs per missile). If we assume the combined missile and warhead reliability to be .8, then the minimum reliable force will be about 100 RVs with one boat on patrol. Two boats on patrol bring this force to about 200 RVs and three boats to about 300 RVs.

Targeting

Although the British do not publicize their targeting for these forces, they must transmit some knowledge of their intentions and capabilities to the Soviets so as to deter them. Thus, British public declarations, vague as they are, should provide clues to their intentions. Evidence for British targeting objectives can be found in official statements on the matter. In 1980, Francis Pym, then minister of defence, stated that "successive United Kingdom Governments have always declined to make public their nuclear targeting policy and plans, or to define precisely what minimum level of destructive capability they judged necessary for deterrence. The Government however thinks it right to make clear that their concept of deterrence is concerned essentially with *posing a potential threat to key aspects of Soviet state power* [emphasis added]."[11]

This key targeting objective can be broken down into two elements. This first is the notion of proportionality. Elsewhere in the document Pym said, "Overwhelming Britain would be a much smaller prize to an aggressor than overwhelming the United States, and a smaller prospective penalty could therefore suffice to tilt his assessment against starting aggression that would risk incurring the penalty."[12] Thus, the British urge the Soviets to perform a cost-benefit analysis before they act. The strategic objective of a British nuclear strike would be to make certain that this cost-benefit calculation never appears to favor the Soviets.

The second element in Britain's targeting objective is an extension of the notion of proportionality by placing it in the context of the geopolitical reality of the U.S.-Soviet competition. Pym goes on to warn the Soviets to consider all costs from this potential British attack. "Indeed, one practical approach to judging how much deterrent power Britain needs is to consider what type and scale of damage the Soviets might think likely to leave them critically handicapped afterwards in continuing confrontation with a relatively unscathed United States."[13] More concretely, how does this translate into specific targets? Is Moscow critically important? Or would other cities suffice? Would attacking a specific set of military targets deter the Soviets?

Targeting for British strategic forces is not static but probably varies with changing capabilities. It is quite possible that SSBNs have a different target set when they are patrolling alone than when two or more are on patrol. The expansion and modernization of British strategic forces through the acquisition of the Trident system will produce another change in targeting. There is no unchanging, absolutely critical Soviet target set that always must be attacked. Rather the British will try to maximize the expected damage to the Soviet value system for any given attack size.

To maximize the expected damage, the attacker needs to consider two factors. First, those targets with the greatest relative density of value should be attacked. Obviously, an attacker would prefer to have the capability of threatening with a 90 percent damage expectancy (DE)[14] a particular target set that contained the Kremlin rather than one that contained a block in downtown Krasnoyarsk or even an airfield adjacent to the city. In this example, the choice is obvious. But, in general, there is no obvious metric by which the value of targets can be ranked. A metric must be imposed somewhat arbitrarily.[15]

Second, a detonating warhead reduces the value of a target set attacked. Subsequent warheads will destroy less value of the same target set because less value remains. Stated another way, there is a declining marginal utility for attacks against a given target set. Thus, there will come a point such that after a certain number of warheads have denotated on Moscow, more value could be destroyed by the next warhead if it were directed at downtown Krasnoyarsk.

To illustrate these aspects of attack planning, consider Figure 1. Curve A is a plot of the value destroyed of some particular target set by the number of attacking RVs—for example, Moscow. Curve B represents an attack against some other target set of lesser value—for example, Krasnoyarsk. In both cases, an attack against the same target base is characterized by declining marginal utility in the attack. That is, each additional warhead does less damage to the target set than the previous warhead until at some point almost no additional damage is done. Furthermore, after some number of warheads have detonated against the more valuable target (Moscow), more damage can be done by subsequent RVs if they are directed against another intact target (Krasnoyarsk).

If, for example, an attacker had a total of X_3 warheads, the attacker could do more overall damage by using X_2 of these to attack target set A and X_1 of these to attack target set B (where $X_1 + X_2 = X_3$) than by devoting the entire attack to target set A. That is because the value of target set B destroyed in going from the origin to point T (by using X_1 warheads) is greater than the value destroyed by using the same number of warheads in going from point U to V (using the same X_1 warheads).

Thus, the targeting question for the British amounts to maximizing the damage to the Soviet value structure with the given British forces.[16] In principle, one can lay out an attack that maximizes the damage to the value structure for the force size and type. But one also must describe the value structure. Moscow is unquestionably the target with the highest concentration of value that could be successfully attacked. Regardless of whether economic, political, or military value (or some combination of these) is considered to be most important, Moscow contains the highest concentration of that value. However, after a certain portion of the value of Moscow were destroyed, the attack would shift to other targets if warheads remained.

Fig. 1—Value of two undefended Soviet target sets destroyed by number of attacking warheads

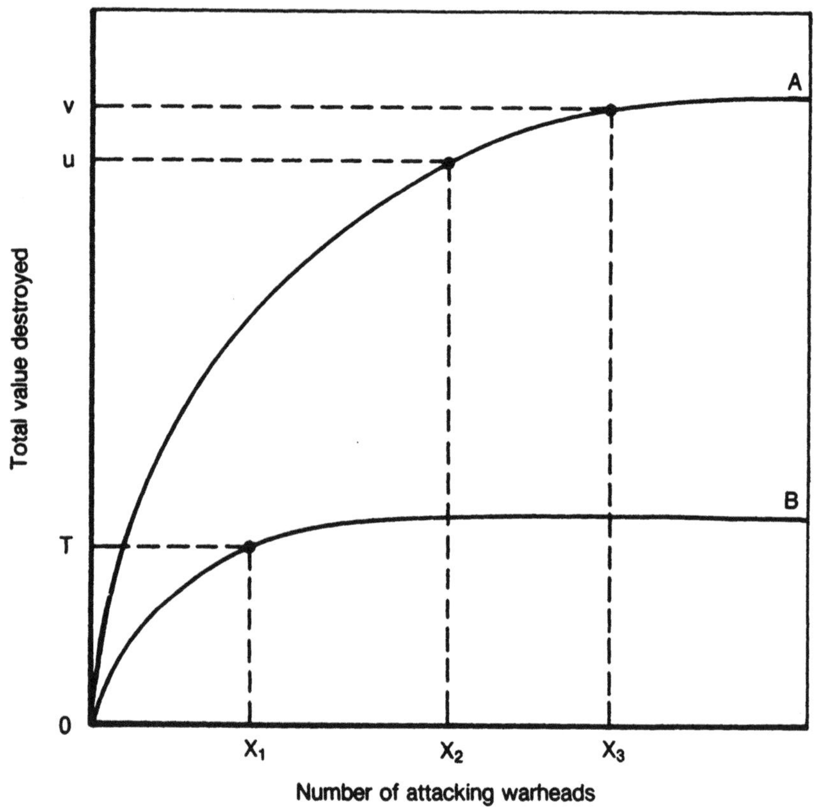

A = results of attacks against undefended Soviet target set (first)
B = results of attacks against undefended Soviet target set (second)
T = total value destroyed for one attacking warhead
U = total value destroyed for two attacking warheads
V = total value destroyed for three attacking warheads

French Strategic Forces

Rationale

In 1961, soon after France exploded its first atomic bomb, General Pierre Gallois articulated a formal rationale for small nuclear forces in *The Balance of Terror.*[17] Gallois believed that thermonuclear weapons had changed the entire nature of military strategy. In the nuclear age, the general said, governments possessing nuclear weapons could practice the politics of deterrence based on thermonuclear "terrorism." Gallois described this form of deterrence, which he called proportional deterrence.

> The thermonuclear force can be proportional to the value of the stake it is defending. Naturally, the value of this stake is fixed by the potential aggressor, and it is on the latter's estimate, however subjective and tentatively known this is, that the nature and power of an effective dissuasion force will depend. It is obvious that in the age of nuclear explosives, the dangers of violence or the threat of violence are so great that the risks run must be constantly compared to the value of the goals violence is intended to achieve.[18]

Proportional deterrence, even today, remains the backbone of French nuclear doctrine. But, with a change in the regional conventional balance between Warsaw Pact forces and NATO forces and a series of local conflicts around the world came a growing French realization that the full-scale nuclear response dictated by proportional deterrence required some modifications.

A 1972 government white paper provided a rationale for tactical nuclear weapons as support for French strategic nuclear forces. Tactic nuclear weapons, which are more limited than strategic forces, were to be used to reinforce the conventional forces dedicated to testing the adversary's intentions and to fire warning shots to convince the adversary of the seriousness of French resolve. After these warning shots had been fired, the French would use their strategic forces.

When Valéry Giscard d'Estaing took office in 1974, he emphasized both the role of tactical nuclear weapons and the necessity for strengthening conventional forces to increase the credibility of the deterrent by providing a greater continuity in the responses available for various situations.[19] In 1976, General Guy Mery extended Giscard's logic and suggested that France might participate in the "forward battle" for Germany as part of the doctrine of "enlarged sanctuarization," which extended French national territory to include contiguous Europe and the Mediterranean basin.[20] This policy suggested that France would be willing to use nuclear weapons in defense of the territory of another country and that an attack on Germany might be tantamount to an attack on France. Nevertheless, Giscard never abandoned the principle of proportional deterrence and insisted on maintaining French nuclear independence.[21]

Present French strategic policy under François Mitterrand also is based on proportional deterrence. If Mitterrand's five-year military plan is followed, France will make significant cuts in the size of its conventional armed forces. The cuts may amount to about 7 percent of the army's personnel during a ten-year period.[22] However, France will maintain roughly the previous levels of spending for nuclear weapons, and, as Mitterrand has approved the construction of the seventh nuclear submarine, the future of both the *force de frappe* and the principle of proportional deterrence seem assured.

French Strategic Posture

In 1963, the French attained a nuclear capability with Mirage IV aircraft carrying a 60-kt nuclear weapon.[23] In 1971, nine S-2 ballistic missiles with a 1,860-mile range were deployed on the plateau d'Albion as the second operational leg of the nuclear triad. The warhead on this first missile was a single fission warhead with a payload of 150 kt. The next year the triad was completed when the French launched their first submarine, the *Redoutable,* equipped with sixteen S-2 missiles. It was followed in short order by the *Terrible* and the *Foudroyant.*

France tested its first thermonuclear device in 1968, but it was not until 1977 that a new missile—the M-20—was made operational. This missile, which had a new single fusion warhead of about 1 megaton, was put on the new submarine commissioned that year, the *Indomptable,* and each old submarine was fitted with the new missiles when it came in dry dock. In 1980, the *Tonnant,* the last of the present fleet of French submarines, was put into service, and nine of the eighteen land-based S-2s were upgraded to S-3s, which are essentially M-20 missiles. In 1983, the other nine S-2s were upgraded.

The French strategic forces are in the midst of modernization and expansion. When the *Inflexible* becomes operational, it will be France's sixth operational SSBN. This new SSBN will carry the M-4 missile, which has a range of more than 2,500 miles, enough to reach most areas in the western Soviet Union from the likely patrol areas of the French strategic fleet. Four of the five remaining boats will be converted from the present M-20 to the M-4 by about 1993.[24] Not only does the M-4 have greater range than the M-20, but it also has six MRV warheads.

By the mid-1990s, another SSBN, carrying the M-5 missile, is expected to be deployed. Not much is known about the M-5, but like the M-4 it is expected to carry six warheads.[25] By this time, the *Redoutable,* the first SSBN deployed, most likely will be retired, leaving the French with a total fleet of six SSBNs.

By the mid-1990s, the land-based IRBM force also is expected to be modernized. The S-3s will be replaced by the new road-mobile SX missile. These missiles will be dispersed in time of crisis, thus reducing their vulnerability. The SX will have a range of about 2,500 miles, enabling it to reach all targets in the western Soviet Union, including Moscow. It is estimated that the French will procure 100 SX missiles.[26]

By the mid-1990s, the Mirage IV aircraft will be retired, removing the air-breathing leg of the French strategic triad. By the early 1990s, the Mirage III and Jaguar will be replaced by Mirage 2000N, carrying the ASMP air-to-surface nuclear missile. But the Mirage 2000N does not have enough range to be used effectively in a strategic role.[27]

French plans beyond the mid-1990s are not known. However, it is assumed that the French will probably construct two more SSBNs, bringing the total to eight, with four always on patrol. In 1980, the Gaullists called for the construction of fifteen SSBNs by the year 2000.[28] However, this would require an ambitious building program. During the 1990s, the four SSBNs built during the 1970s will probably be retired because they will have become obsolescent.[29] Thus, the expansion of the force beyond six operational boats would require considerable effort. The French would have to replace the retired boats before they could further expand the fleet.

Targeting

Independent proportional deterrence seeks to deter aggression by threatening to make the aggressor's losses as a result of the aggression greater than its gains. Thus, it is not necessary for the French to match the arsenal of the Soviets. Rather, they must have only a sufficient nuclear force to threaten a high enough level of damage to the Soviets under any conceivable circumstances. Furthermore, the ability to inflict this damage must not rely on any other country. The French, like the British, have been vague about the requirements for actually implementing this doctrine. With changes in forces, the exact target set undoubtedly has changed.

French spokesmen in the 1970s publicly suggested that France had an "anti-cities doctrine," but they did not elaborate the nature of that doctrine. In fact, the French considered this vagueness a virtue because it increased Soviet uncertainty, which, in turn, it is claimed, increased deterrence. The "anti-cities" targeting doctrine changed in the 1980s.[30] Whereas the earlier policy suggested that the French strategic forces could be aimed at urban populations, the new, "enlarged anti-cities" policy suggested that the Soviet economy and industrial recovery potential would be targeted. This new doctrine probably resulted from both increased capabilities and new strategic goals.

The introduction in the 1990s of the M-4 missile, with its six warheads and MRV front end, will give the French many more RVs for their planning. French targeters will use these forces to present the greatest possible threat to the value base of the Soviet Union. Moscow undoubtedly would be attacked first. But, when the marginal return from attacking some particular target set other than Moscow is greater than the return from continuing the attack against Moscow, then these other targets would be attacked. Thus, an increasing force size would lead to an expanded attack potential because it would be irrational to bounce the rubble in Moscow when other lucrative targets could be attacked.

Soviet BMD and British/French Strategic Force Effectiveness

Although President Ronald Reagan's speech on March 23, 1983, focused national and world attention on strategic defense, the Soviets have long had an interest in it. They have devoted considerable effort to developing a ballistic missile defense system. According to published reports, they are developing what is called the ABM-X-3, which would have a reasonably sophisticated SH-04 exoatmospheric interceptor and an SH-08 endoatmospheric interceptor.[31] In addition, they have been deploying a large number of phased-array radars that could be netted into the ABM-X-3 system to provide critical early warning to the system.

The Soviets also have been active in the research and development of space-based systems, but these components of a BMD system are not as developed as are the land-based components. The deployment of a system in the near term probably would begin with the land-based component, perhaps followed by a space-based component some time later.[32]

The capability of such a system would be impossible to assess accurately because such an assessment would have to be made while the system was still under construction. In fact, even the goal of such a system could not be known. It could be designed to reduce damage from a U.S. attack. However, this could be a long-term objective. In the interim, a less-than-adequate system against a U.S. attack could offer substantial protection against smaller, third-party attacks.

If the Soviets began to deploy an extensive defense, the British and the French very likely would have to prepare for a worst-case scenario. They would have to assume that the BMD was going to work effectively, if not immediately, then sometime in the near to middle term. However, a Soviet BMD system would not appear fully deployed overnight. Presumably, the British and the French would see the beginnings of such a deployment and would have time to respond. How might they respond in order to maintain deterrence of the Soviets? Do they have reasonable and affordable response options that would enable them to maintain deterrence?

Response Options

If the deployment of a Soviet BMD threatened to undermine the effectiveness of the British and French strategic forces, both nations would have several options available to counter this threat. Either ally would select a response after a thorough analysis of the situation. The response would depend on many factors, including the country's own technical strengths and capabilities as well as an estimate of the capabilities of the Soviet BMD. Let us consider some of these options.

Cruise Missiles

One possible response option available to both countries would avoid the Soviet BMD by abandoning ballistic missiles and moving to a cruise missile threat. For the British, the viability of this option would depend on the cooperation of the United States, especially with advanced-technology (e.g., Stealth) cruise missiles, and the sunk costs that already had gone into the development and production of British SSBNs and warheads. If the sale of the D-5 missiles had not reached a point of no return, then perhaps the British could purchase U.S. ground-launched cruise missiles (GLCMs) or sea-launched cruise missiles (SLCMs) instead of the D-5s. The GLCMs could be land based; the SLCMs could be carried either by British submarines designed exclusively to carry such missiles or by the fleet of nuclear-powered attack submarines (SSNs), perhaps expanded somewhat. The British considered this option when they analyzed modernization options for the strategic forces in the 1970s.[33]

When this modernization decision was being made in the early 1970s, many argued for acquiring cruise missiles instead of SLBMs on the grounds that cruise missiles would be cheaper. Consideration was given to acquiring some additional GLCMs that the United States was about to deploy in the UK. The additional GLCMs would have been under UK control and deployed throughout the country.

The vulnerability of the GLCMs to a Soviet surprise attack presented a major problem. If these missiles were not dispersed, but were still in their garrisons, the Soviets could destroy them. Although a bolt from the blue seemed quite unlikely, the deterrent force would not be as credible as the less-vulnerable SSBN force. The purpose of an independent force was to deter the Soviets under all circumstances. Furthermore, in a time of heightened tensions, these missiles would have to be dispersed in order to be survivable. However, dispersal could add to the tensions. For these reasons, the GLCM option was discarded.

More serious consideration was given to acquiring a SLCM force. It was argued that this force would be just as effective as a ballistic missile force but cheaper. However, this argument was not persuasive. Although each cruise missile would be cheaper than a ballistic missile, many cruise missiles would be required to equal the potential destructive power of one MIRVed SLBM. Aside from the Moscow antiballistic missile (ABM) system, submarine-launched ballistic missiles (SLBMs) would get a "free ride," whereas the SLCMs would have to penetrate the Soviet air defenses. In addition, a large submarine fleet would still be needed to carry these missiles, perhaps larger than that needed for Trident, for equal effectiveness. Because submarines constitute the major cost in the submarine-missile system, this increased requirement for submarine carriers of SLCMs would make an SLCM system more expensive than an SLBM system.

Alternatively, these cruise missiles could be placed on board existing SSNs. But, during a crisis, this SSN fleet would be deployed in an antisubmarine warfare (ASW) role. To be effective in this role, they would have to be deployed forward and aggressively pursue their quarry. Clearly, this type of operation would be inconsistent with the requirement to hold the strategic deterrent force in reserve. Thus, SLCMs could not be effectively deployed aboard the SSN fleet.

These arguments were persuasive in 1979—in the absence of a full-scale Soviet BMD deployment. If the Soviets began deploying a BMD system, the cruise missile option again would be considered. Since the British made the decision in the late 1970s, the cost argument has not changed. They would have to acquire a reasonably large and expensive submarine fleet to carry the SLCM missiles. However, during the last few years, the effectiveness side of the cost-effectiveness equation may have deteriorated.

Cruise missiles do not have a free ride to their targets. They are a slow-moving target, able to penetrate the extensive Soviet air defense network by presenting a small signature and low contrast with the ground clutter. However, their penetration depends on a successful measure-countermeasure competition. As the Soviets develop and deploy Airborne Warning and Control System (AWACS) and look-down, shoot-down capabilities for their aircraft, cruise missiles will bcome more vulnerable. In the past few years, Soviet capabilities in this area have improved in response to the growing U.S. cruise missile threat. Further Soviet improvements seem likely.

The United States has countered such improvements with various Stealth technologies and with electronic countermeasures (ECM) that further reduce the missile signature. In addition to these measures, the United States has other options open to increase the penetration probability of its cruise missiles, including the suppression of the Soviet air defenses with SLBMs or the saturation of these defenses with large attacks of cruise missiles in relatively restricted areas. The British could neither suppress nor saturate an air defense.

As a result, if the British acquired cruise missiles for their strategic deterrent, they would enter a high-technology measure-countermeasure competition. This competition could turn out to be expensive with upgrades required to maintain confidence in the penetration probability of the British force. Even then, the British could not have a high degree of confidence that they could continue to stay ahead in the competition.

Perhaps with continued U.S. aid, the British could maintain an effective cruise missile force, but that force would depend on the United States to maintain its effectiveness. This situation would undermine the very reason for the force. Furthermore, Britain could not be sure that the United States always would want to share critical penetration information.

The British once considered and rejected the cruise missile option. In anticipation of the deployment of an extensive Soviet BMD, they would probably again consider cruise missiles as a strategic deterrent. However, for the foregoing reasons, this option is likely to be rejected again.

The French, with their sophisticated aircraft industry, probably could produce the propulsion and avionics required for cruise missile development without U.S. help. The terrain contour mapping (TERCOM) required for guidance might prove more difficult but would probably not be beyond French capability.[34] Nevertheless, cruise missiles are as unlikely to provide a viable alternative for the French deterrent force as for the British and for many of the same reasons.

The French have maintained a triad of strategic forces. However, the air-breathing and land-based ballistic missile legs of their triad have become more vulnerable to a Soviet surprise attack. By the mid-1990s, their Mirage 2000 aircraft and SX missiles will depend upon disperson to survive a Soviet attack. France must be able to deter a bolt from the blue, however unlikely, if it is to maintain the credibility of its nuclear forces. Thus, because the submarines on patrol would survive a Soviet surprise attack, this leg of the triad must be maintained, and GLCMs are not a viable option.

The French could place SLCMs aboard their submarines, but this option would have the same problems for them as for the British. SLCMs probably would not be cost effective because the requirement for more submarines (which could not be SSNs) would increase the cost. Furthermore, the continued effectiveness of French cruise missiles against Soviet air defenses would depend on a sophisticated measure-countermeasure competition with the Soviets. Moreover, France is even less likely than Britain to receive cutting-edge technology in this realm.

Another difficulty for the French, one not completely shared by the British, stems from the sunk costs in their SSBN fleet. Whereas the British will modernize their entire deterrent force in a short time, the French modernization program is a continual one, with new boats being added to the fleet one at a time. The newest boats are expected to remain in service for perhaps twenty-five years. It would be difficult for the French to change direction and abandon this SSBN fleet for cruise missiles.

Unless cruise missiles appear to be overwhelmingly attractive, the French seem more likely to improve their SSBN fleet than to switch to cruise missiles. Thus, cruise missiles do not seem to be a viable response option to a potentially effective Soviet BMD for either the British or the French.

Change Targeting

The British and the French share a targeting aim: to maximize the destruction of the Soviet value system with available forces. Because Moscow is the Soviet Union's most valuable asset, both British and French targeting probably focus particularly on Moscow. But, because it is the most valuable, the Soviets would defend it most heavily. If the Soviets reduced the potential damage to Moscow to some minimal level, are there other attractive targeting options that do not include Moscow?

No defense can promise to eliminate all damage from an attack. The purpose of a defense is to eliminate the asymmetric situation in which disproportionate value can be destroyed for any given attack size. For example, if a given area of Moscow is ten times as valuable as a given area of Kiev, then the Soviets would deploy a defense that would require an attack size against the area of Moscow to be ten times as large as an attack size against an area in Kiev to achieve the same DE. Thus, a defense would eliminate the disproportionate value destroyed for a particular attack size.

In principle, the effective defense would not be absolutely leakproof but would make the expected damage against the overall value scheme constant for a given attack size and considerably less than the expected damage from the same attack against an undefended value scheme. Thus, in principle, the deployment of an extensive Soviet defensive system would result in the elimination of alternative targeting schemes. Any particular attack would be charged a particular price to destroy a given target value. No one attack would be better than any other attack in terms of aggregate value destroyed.[35]

In fact as well as in principle the Soviets would try to defend their value base in the manner just described. A successful defense would protect strongly the most valuable areas of the Soviet Union, such as Moscow, and would reduce the potential damage to the Soviet value base for a given attack size. No change in targeting can overcome this. However, there could be some advantage in combining and coordinating British and French attacks. Assuming that the British and the French are interested in maximizing the damage done to the Soviet value system, the situation with Soviet defenses is now considerably more complicated.

Consider the notional situation depicted in Figure 2. The top curve in Figure 2 is the relationship between attacking RVs and target value destroyed for target set A as shown in Figure 1. A defense would degrade this offensive capability.

A different notional curve (curve C) illustrates what might happen with defense. From the origin to point T on curve C, the defense works effectively, and few RVs leak through. Thus, an attack of size X_1 produces little damage. Then, the defense degrades, and from points T to U on curve C the attack becomes more effective and the rate at which the target set is destroyed increases more rapidly. At point U on curve C, the defense is exhausted, and the rate at which the target set is destroyed increases even more rapidly after point U.

Thus, in contast to the undefended situation in which there was always a decreasing marginal utility to any attack, now there may be a rapidly increasing marginal utility to increasing attack size. If, for example, X_1 was the size of the British attacking force, then it could damage only about 23 percent of the target base represented in Figure 2. However, if point X_2 represented the combined total of the British and the French attacking forces, then the allies would have considerable incentive to combine their forces under joint targeting arrangements because the combined forces could destroy about 73 percent of the target value in Figure 2.

This is only notional. A real analysis would have to consider the defense as well as the offenses. However, defenses may well produce a situation of increasing marginal utility for additional attack size. In that case, coordinating forces may well be in the interest of both the UK and France.

Unfortunately, the coordination of forces would be quite difficult. To achieve maximum targeting effectiveness, the type of message conveyed to the submarines of both forces ordering an attack must be more complicated than a simple go–no go. At the very least, each SSBN must have several different target sets to maximize targeting effectiveness for different numbers of attacking submarines because it could not be known in advance how many SSBNs would be available for an attack. The Soviets could destroy some SSBNs in port and others with antisubmarine

Fig. 2—Value of one defended Soviet target set destroyed
by number of attacking warheads

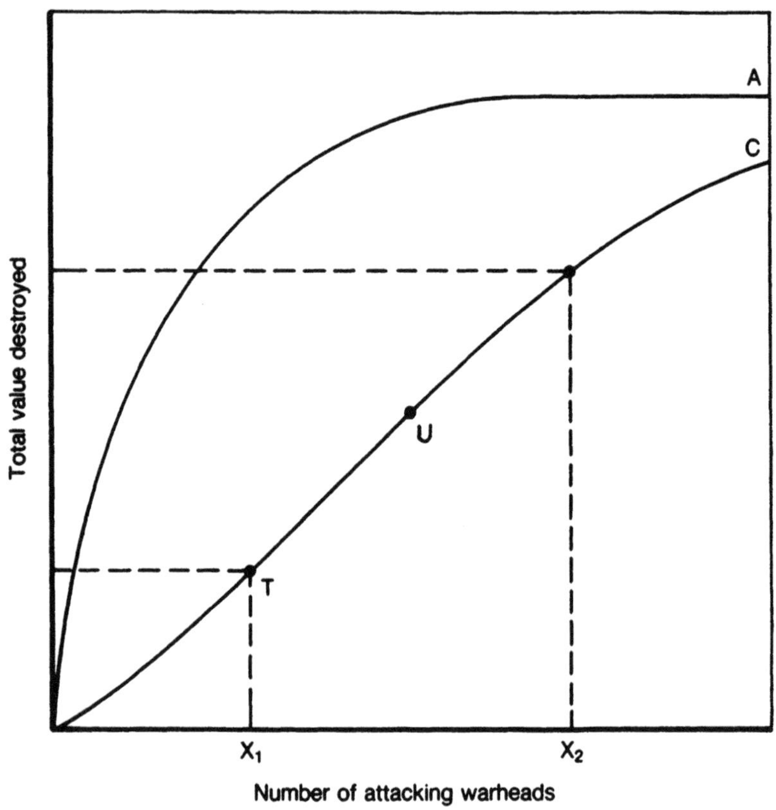

A = results of attacks against undefended Soviet target set (first)
B = results of attacks against undefended Soviet target set (second)
C = results of attacks against defended Soviet target set (first)
D = results of attacks against defended Soviet target set (second)
T = total value destroyed for one attacking warhead
U = total value destroyed for two attacking warheads
V = total value destroyed for three attacking warheads

warfare means. For France and Britain to maximize the effectiveness of their attack, they would have to have some way of knowing how many SSBNs had survived and of communicating that information to the surviving SSBNs to direct their targeting. Otherwise, all SSBNs would be targeted inefficiently, and targeting effectiveness would be reduced.

These changes in targeting would be difficult for either country to convey to their SSBNs because of the vulnerability of their command, control, communications, and intelligence (C^3I) systems to Soviet attacks. However, coordinating an attack between the two countries would be even more difficult. Both countries would have to agree on the appropriate moment for the counterattack. Coordinated preplanning would be needed to spell out options. Then, France and Britain each would have to assess accurately its own situation and pass that information to the other country so that a joint decision could be made on the appropriate option. Either could have difficulty making such an assessment. Relaying the information also could be difficult because command, control, and communications (C^3) systems might be destroyed. But, more importantly, this information would be critical to their survival, and each country might not be willing to share it with the other country in a crisis, even if both had agreed to do so beforehand.

Although alternative targeting schemes alone would not increase the effectiveness of each separate attacking force and combining forces would prove difficult, both the British and the French probably will change their individual targeting against a defense. In the undefended case, the declining marginal utility to attacking any given target leads the attacker to focus on more than one target to maximize the damage done to the target base (this was illustrated in Figure 1). In the defended case this may not be true.

Figure 3 shows the same case of an attack against two defended target sets. Underneath the original attacks against undefended target sets (curves A and B) are notional examples with defenses (curves C and D). X_3 represents the total number of attacking RVs. Here we see that if, for example, the attacker were to use X_2 of the total number of RVs to attack the more valuable target set and X_1 RVs to attack the less valuable target set, the total value destroyed would be less than the value destroyed by using all RVs (X_3) to attack the more valuable target set (curve C). This conclusion is based on the fact that for the example considered here, the portion of curve C between X_2 and X_3 is steeper than the corresponding portion of curve D between the origin and X_1. Thus, it would be better for an attacker to concentrate an attack on the more valuable target set.

This is only a notional example, but it illustrates a point. Defenses might make an attacker concentrate an attack against a smaller number of high-value targets because there is likely to be an increasing marginal utility in an attack against any given target. In the case of small attack sizes, such as the UK and France will have in the mid-1990s, this could well mean concentrating the entire attack against Moscow.

Fig. 3—Value of two defended Soviet target sets destroyed by number of attacking warheads

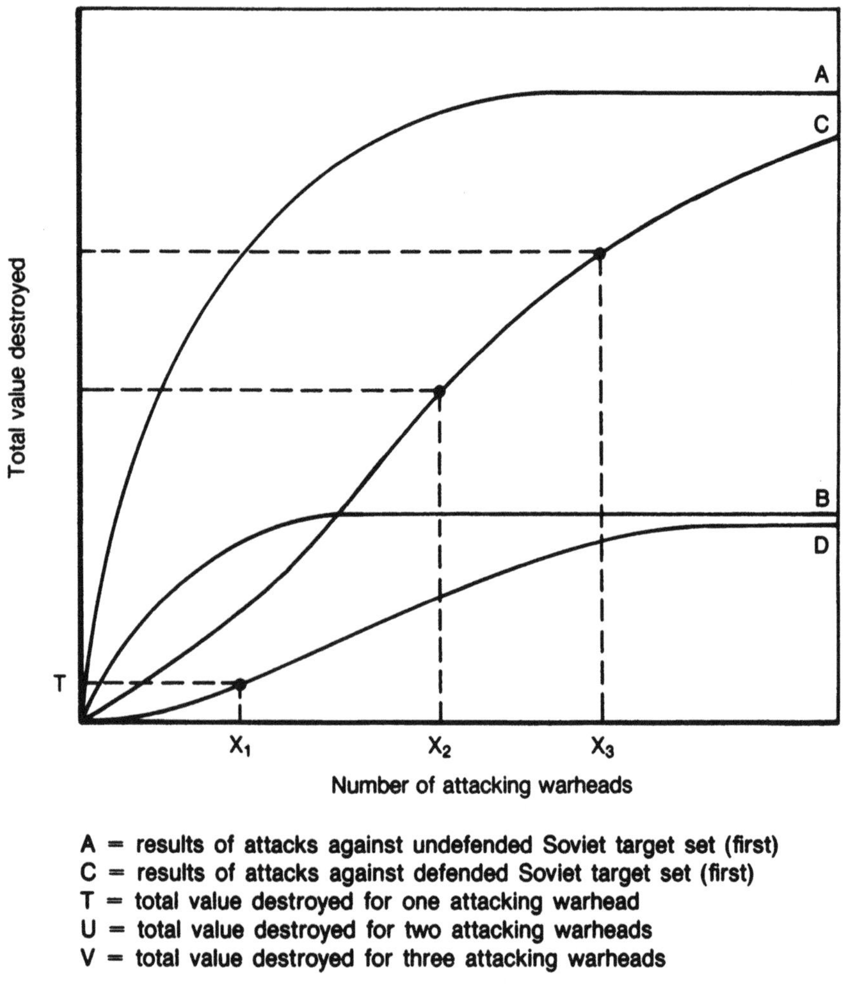

A = results of attacks against undefended Soviet target set (first)
C = results of attacks against defended Soviet target set (first)
T = total value destroyed for one attacking warhead
U = total value destroyed for two attacking warheads
V = total value destroyed for three attacking warheads

Defense Suppression

A third option to consider would be defense suppression. If defense effectiveness could be seriously undermined by a small attack, then the remaining RVs could be directed against undefended targets, such as Moscow. If the Soviets deployed a BMD system, it would be likely to have large, phased-array, early-warning radars. These radars would track incoming RVs and would hand off that information to other radars in the system. The early-warning radars would be very soft targets; they could not be appreciably hardened.

If these radars were not defended, they would present easy targets. However, the Soviets almost certainly would defend them.[36] The radars would be assigned a certain value and in principle would be defended in such a way that any attack against them would not be more profitable than any other attack against a different target set. Thus, there would be no disproportionate value destroyed in these attacks.

In addition, attacks against these radars would present special problems for any attacker, but especially an attacker with a relatively small force. The problem would stem from the footprint constraints of the warheads. The smallest attack that could be directed against a particular radar would be the full complement of warheads contained on one missile. But, this attack is both too small and too large at the same time.

The attack is too large because the DE to one soft radar from six to eight attacking warheads would be much higher than would be required. Perhaps one or two warheads would suffice. However, each British D-5 will contain eight, and each French M-4 will contain six MIRVed warheads, most of which would not be used effectively because many of the Soviet radars would be located in isolated areas on the periphery of the Soviet Union. It is likely that there would be no other high-value targets close enough that they could be attacked by other RVs from the same missile.

Yet, this attack is too small because in spite of a rather large DE from six to eight attacking warheads against each of the radars, there is a reasonably high probability that the entire missile might malfunction. Then, none of the warheads would reach the target. To guard against this possibility, radars could be cross-targeted from different missiles. But this would introduce even greater inefficiencies in the employment of the force.

Thus, it is difficult to plan an optimal attack against a remote, very high-value target. These kinds of attacks can be inefficient and can require the use of a large number of RVs to ensure success. A small attacker may not be able to afford this luxury.

There is yet another problem. Considerable redundancy would be built into the defense system. Even if a particular early warning radar could be destroyed, it would be very difficult to estimate the effect of that loss on the overall performance of the system. The lack of a handoff reduces, but does not totally destroy, the effectiveness of shorter-range radars. Thus, in spite of an attempt at suppression (at considerable cost to the attacker in terms of RVs expended), the BMD system may operate with very little degradation.

Decoys and Penetration Aids

The use of decoys or penetration aids (penaids) offers yet another possibility for increasing the effectiveness of an attack. The relative effectiveness of either approach would depend on the configuration of the defense. Decoys can be particularly cost effective during the exoatmospheric portion of the RV flight. The simplest decoys are balloons that the missile carries deflated in its front end in the boost phase of its flight. After separation of the front end of the missile from its booster, these balloons are inflated to the appropriate size outside of the atmosphere. If the surface is coated with a reflective material, these balloons are indistinguishable to radar from real RVs. In the exoatmospheric portion of their flight, the balloons travel at the same velocity as the real RVs. Because the balloons are very lightweight, many can be carried in place of each RV to produce many false targets.

There are methods of discrimination between real RVs and decoys outside the atmosphere. Infrared sensors can discriminate because the thermal properties of RVs and decoys differ. Or more exotic techniques could be used, such as radiating all objects with a beam of subatomic particles and observing different responses. But these techniques are not yet well developed.

When these objects hit the atmosphere, the problem of discriminating lightweight balloons from real RVs disappears because friction with the atmosphere drastically alters the trajectories of the balloons relative to the real RVs. There is no simple way to overcome this distinction because any decoy that had the weight of a real RV might just as well be an RV.

A relatively complicated procedure for masking decoys in the endoatmosphere is to equip them with rocket thrusters to increase their velocity through the atmosphere to that of the RVs. However, other techniques can be used to discriminate between the decoy and the RV. For example, their wakes can be examined as they move through the atmosphere.

The effectiveness of any of these techniques will depend on the nature of the defense. If the defense were to contain a space-based component that depends on simple radars with no other discriminators, then a simple balloon-type decoy could be very effective, at least in the exoatmospheric portion of the flight of the RVs. However, if the defense were to contain more sophisticated discriminators, then such simple decoys may not be effective. In any case, simple balloon decoys are not likely to be very effective in the endoatmospheric portion of flight against the ground-based portion of the defense.

Penetration aids are similar to decoys in that they are intended to increase the penetration probability of the RVs. These can consist of simple things such as chaff released during the flight of the missile to confuse the ground-based radars at critical times. During the exoatmospheric portion of the flight of the missile, the chaff cloud will travel with the RVs, thereby effectively hiding them from radar. The atmosphere will slow the chaff cloud, but it still provides the reflective background to confuse the ground-based portion of the defense.

There are other, more sophisticated penetration aids. Various types of ECM can be carried aboard the bus to confuse the defenses. Or the RVs can be made to change their course during their flight so that they cannot be easily tracked.

Either decoys or penaids could be quite effective in increasing the penetration probability of an attacking force, depending on the type and configuration of the defense. Thanks to the substantial capacity of the D-5 missile, a large package of penaids or decoys could be carried. Thus, the British would certainly explore this avenue. However, this high-technology solution to the problem has been tried before. The experience with the Chevaline was an unhappy one. The program was quite expensive (about £1 billion) and might have generated controversy except for the secrecy surrounding it.[37] In addition to the potential expense of employing penaids, this approach would take the British into measure-countermeasure competition with the Soviets.

The front end of the French M-4 missile is reported to contain chaff and ECM, which may increase the penetration probability of the warheads.[38] The French also are developing this technology as part of their entire nuclear capability.[39] The effectiveness of this system cannot be determined, but French experience in this area would probably be developed further in the face of a Soviet BMD deployment.

Increase Attack Size

The final alternative is the simplest and perhaps the most cost effective, at least for the British. If the Soviets move to deploy a BMD, the British could increase their attack size substantially to increase the assurance of a sufficient level of damage to produce deterrence. Given that the SSBN-ballistic missile system has not yet been deployed, the attack size probably could be increased at minimal cost by designing the front end of the D-5 to mazimize the number of RVs carried. The D-5 is a much larger missile than the A-3 or the C-4 and has a much greater potential carrying capacity. It has been reported that the D-5 can carry fourteen 150-kt warheads.[40] If the warhead yield were reduced, the D-5 could probably carry about twenty 100-kt warheads. This would represent an increase of 2.5 times the attack size of the eight 100-kt warheads planned for the C-4.

With this twenty-warhead missile, the minimal attack size with only one boat on patrol (assuming a reliability of .8) would be about 250 reliable RVs. More likely than the bolt-from-the-blue scenario is the situation in which the British would have two boats on patrol after increasing tensions. These two boats would provide the British with about 500 reliable RVs. In fact, by increasing the SSBN fleet size from four to five, the British could be assured of having at least two boats on patrol at all times. A further increase in the fleet size to six would assure a minimum RV attack of about 750 because three boats could always be on patrol.

Thus, by increasing fractionation (number of warheads per launcher) and expanding the fleet, the British could increase their attacking force size by a factor of about 7.5 over their planned expansion and about twenty times their present reliable, minimum attack size of about 40 RVs. If we assume linearly increasing costs for acquisition, then acquiring a six-boat fleet would cost about 1.5 times as much as acquiring a four-boat fleet or perhaps somewhat more if we allow for the expense of additional warheads and guidance. Thus, 7.5 times the planned, expanded force size could be procured for about 1.5 times the present projected expenditures. Because the total procurement cost of the Trident as planned is expected to be about 9 percent of the total defense budget between 1980 and 1995, an increase to 14 or 15 percent of the defense budget would buy 7.5 times the previously planned force size.[41]

These calculations are only illustrative. The size of the attack necessary to produce deterrence would depend on a thorough analysis of the defense system and cannot be determined here. A minimum attack of 750 reliable RVs might not suffice to deter. Perhaps decoys and/or penaids could be used in conjunction with an increase in the attack force size to increase its effectiveness. The point is that with the capacity of the D-5 missile, the British have a relatively cheap option for substantially increasing their threat to the Soviets. Reducing the potential effectiveness of this threat to an acceptable level would not be easy for the Soviets, especially in the next fifteen years.

The French cannot expand their force size as rapidly or as cheaply as can the British. The M-4 already has been designed with six RVs per missile. Little is known about the design of the follow-on missile—the M-5. It is reported to have a slightly longer range than the M-4, probably the result of increased throw weight of the missile. If the French were to devote the increased throw weight to increasing the number of warheads contained in each missile instead of to increasing the range, they probably would generate some potential for expansion of the force size.

Assuming that the next two French boats contained the M-5 with ten warheads per missile (instead of six) and sixteen missiles per SSBN, then the mid-1990s fleet of eight SSBNs with four always on patrol would have a minimum reliable force of about 350 RVs (i.e, three boats with sixteen missiles and six RVs per missile and one boat with sixteen missiles and ten RVs per missile), which are one-half of what the modernized, expanded British force could be like in the mid-1990s. Of course, this large discrepancy in potential force size results from the considerable throw weight of the D-5 missile and the assumption that a front end could be designed containing twenty RVs. Given that the French SSBNs and the M-4 (and perhaps M-5) missile already have been designed and the capacity of the French missiles is so much smaller than that of the British, the French do not have as much flexibility.

Thus, for the mid-1990s, the French are not likely to have many options for dramatically increasing their force size over what they already have planned. Furthermore, there are no promising options for dramatically increasing the effectiveness of these forces in that period. The question is whether this force would deter the Soviets.

Predictions beyond the year 2000 cannot be made with any certainty. New technologies could be developed, and the alternatives considered here could become promising. Or different alternatives could be developed. However, French options appear to be rather limited. Presently, the French are emphasizing the development of their nuclear forces. They are devoting about 30 percent of their defense budget to their nuclear forces, with about two-thirds of that concentrated on their strategic forces.[42] Their conventional forces already are being starved. A further increase in the proportion of their defense budget devoted to nuclear forces might be resisted. However, independent nuclear forces have been the backbone of French security policy for more than twenty years and enjoy great public support. It seems unlikely that the French will give them up easily.

Conclusions

The British and U.S. governments have signed a memorandum of understanding outlining the general areas of participation and cooperation by the two countries in the development of BMD systems. Yet the agreement should not interpreted as complete support by the British for the Strategic Defense Initiative.

Commercial and technological benefits have been major factors in the British decision to cooperate. In addition, the logic of the "second nuclear decision center" also has played a role. The British hesitate to express public concern about the possible effect of BMD on their own independent nuclear forces because they do not like to accentuate the independent role of these forces. Nevertheless, at least privately within the government, the British must be concerned about a possible Soviet BMD, its effect on their strategic nuclear forces, and possible British response options.

The French stress the independence of their strategic forces and so do not hesitate to voice their concern about BMD. They have been more critical than the British of the U.S. SDI and have not entered into any formal cooperative arrangement with the U.S. government. Publicly, they express confidence that they will be able to maintain the effectiveness of their strategic forces. Privately, they, like the British, must be concerned.

The USSR has not begun to deploy a countrywide BMD system; thus, no one can determine how big the problem would be for Britain or France. But both countries are scheduled to greatly expand their strategic nuclear forces during the next ten to fifteen years, when they modernize these forces. Furthermore, both have options beyond this planned expansion to increase the effectiveness of their strategic forces against a Soviet BMD.

In the face of a Soviet deployment of a BMD system, the British probably could substantially increase the size of their threat by increasing the number of RVs carried by the D-5 missile in the modernized force. The French do not have the flexibility to increase their threat size as quickly, but they are known to be concentrating on developing decoys and penetration aids to increase the penetration of a Soviet BMD. Thus, in light of these response options, providing an effective national defense against these two nuclear powers would be a formidable task for the Soviets.

Notes

I would like to acknowledge the useful comments by my Rand colleagues James A. Thomson and William Martel. Of course, I am solely responsible for all errors of fact and interpretation.

1. By European, I mean the NATO countries of Western Europe. Of course, attitudes differ among countries as well as within each country.

2. For a more complete discussion of the British rationale for nuclear weapons, see Christopher J. Bowie and Alan Platt, *British Nuclear Policymaking,* R-3085-AF (Santa Monica, Calif.: Rand Corporation, January 1984).

3. Lawrence Freedman, *Britain and Nuclear Weapons* (London: Macmillan, 1980), p. 5.

4. Ibid., p. 120.

5. Ibid., p. 5.

6. This is stated explicitly in the *Second Report from the Expenditure Committee Session 1975–1976, Defence Minutes of Evidence,* 13 May 1975. Memorandum by the Minister of Defence (SCOE 73/1), 9 May 1975 (London: HMSO), paragraphs 32–34. A further major rationale for the force is its political contribution to the British position in international politics, alliance politics, and intra-European politics. During a time of declining influence in international affairs and a loss of prestige resulting from the loss of its colonies, Britain at least can appear to maintain great power status and a prominent position in international politics—admission to the "top table"—by possessing nuclear weapons. Britain, for example, must be consulted on arms control negotiations. Within Europe, the British can speak with more authority than the economically stronger Germans at NATO Nuclear Planning Group meetings. With nuclear weapons, Britain can deal with France as an equal and does not have to defer to its old rival on the Continent. These political rationales, although simply stated, should not be underestimated. They are powerful motivating forces and very important reasons for the continuing existence of the British strategic nuclear forces.

7. International Institute for Strategic Studies, *The Military Balance, 1984–1985* (London: International Institute for Strategic Studies, 1984).

8. John Roper, "Developments in Ballistic Missile Defense," *The World Today* (May 1985):92–95.

9. *The United Kingdom Trident Programme,* Defence Open Government Document 82/1, Ministry of Defence, March 1982.

10. Placing more than eight RVs on the missile would penalize the United States under the SALT II constraints because all U.S. D-5s would be counted as having this larger number of RVs.

11. *The Future United Kingdom Strategic Nuclear Deterrent Force,* Defence Open Government Document 80/23, Ministry of Defence, July 1980, p. 6.

12. Ibid., p. 5.

13. Ibid.

14. The damage expectancy of a nuclear weapon against a target is a measure of the expected value of the target that will be destroyed by that weapon.

15. For a discussion of British nuclear targeting, see Lawrence Freedman, "British Nuclear Targeting," *Defense Analysis* 1, no. 2 (1985):81–99.

16. Finding a common metric for value is somewhat arbitrary. In the case of two urban areas, population or industrial floor space can be used. But this woud not account for political or military value. These and other types of values have to be defined differently and related to economic value.

17. Pierre Gallois, *The Balance of Terror* (Boston: Houghton Mifflin, 1961).

18. Ibid., p. 138. Less explicit but even more important was the rationale used by de Gaulle. Although accepting Gallois' arguments for the most part, de Gaulle was more interested in using the strategic forces to further his political ends: restoration of French grandeur and reunification of the two halves of Europe. To de Gaulle, the most important consequence of the possession of nuclear weapons and the doctrine of proportional deterrence was the independence that they gave France. De Gaulle felt that France's status in the world had diminished with the loss of its overseas colonies, the defeat in 1940, the humiliation at Suez, and the revival of West Germany as a military and economic power in Europe. Furthermore, he resented the ongoing special relationship between the United States and Britain. With its own nuclear weapons, France no longer had to depend on the United States for its security but could aspire to a global role and equal status with the United States and Britain and a superordinate status with regard to West Germany. In addition, France could be the most important country in a unified and united Europe—most important because it could provide the means for the defense of Europe against the Soviet Union, although the details of this plan were never quite clear. See Wilfrid L. Kohl, *French Nuclear Diplomacy* (Princeton, N.J.: Princeton University Press, 1971).

19. See Valéry Giscard d'Estaing, "Address," *Survival* (September/October 1976):228.

20. See Guy Mery, "French Defense Policy," *Survival* (September/October 1976):226.

21. For a discussion of the evolving French debate, see David Yost, "The French Defense Debate," *Survival* (January/February 1981):19–28.

22. Jacquelyn K. Davis, "France Debates Its Defense Policy," *Air Force Magazine* (September 1984):164–173.

23. See Kohl, *French Nuclear Diplomacy*, p. 180ff.

24. *Armed Forces Journal* (November 1984):58–62.

25. *Defense and Foreign Affairs* (May 1985):9–13.

26. Ibid. Although the French continue to project the deployment of the SX, funds for it have not been included in the 1984–1988 Defense Programming Law. See U.S. Department of State, *France: Implications of the 1984–1988 Defense Programming Law*, Report 710-AR (Washington, D.C.: Department of State, Bureau of Intelligence and Research, November 1, 1983).

27. *Defense and Foreign Affairs* (May 1985):9–13.

28. Yost, "The French Defense Debate," pp. 19–28.

29. The reactor and diving system of an SSBN usually requires increasingly expensive repairs after about twenty years. In addition, the earliest French boats were noisy and thus somewhat vulnerable to Soviet ASW.

30. For a discussion of French targeting, see David Yost, *France's Deterrent Posture and Security in Europe: Part I: Capabilities and Doctrine* (London: International Institute for Strategic Studies, Adelphi Papers no. 194, Winter 1984/1985), p. 15ff.

31. Roper, "Developments in Ballistic Missile Defense."

32. For a general discussion of potential Soviet BMD capabilities, see Sayre Stevens, "The Soviet BMD Program," in Aston B. Carter and David N. Schwartz,

eds., *Ballistic Missile Defense* (Washington, D.C.: The Brookings Institution, 1984), pp. 182–220.

33. For a discussion of this decision, see Freedman, *Britain and Nuclear Weapons,* pp. 69–78.

34. The French have announced a plan to launch a military satellite in the early 1990s. This satellite could possibly be used for the mapping task. See *Aviation Week* (23 December 1985):59.

35. This is an oversimplification of the offense-defense competition. Effective defenses would be layered, and the effectiveness of each layer would depend on many factors. In particular, to charge a constant price, the defender would have to know where each warhead was directed.

36. In addition to blast overpressure, these radars could be vulnerable to electromagnetic pulse (EMP). A single nuclear detonation at high altitude could disable them. However, the effects of EMP are not well understood.

37. For a discussion of the Chevaline program, see Freedman, *Britain and Nuclear Weapons,* pp. 41–51.

38. Lawrence Freedman, "The Small Nuclear Powers," in Carter and Schwartz, *Ballistic Missile Defense,* p. 268.

39. Jean deGalard, "New Approaches for French Nuclear Force," *Defense Weekly* (December 14, 1985):1298.

40. D. Douglas Dalgleish and Larry Schweikart, *Trident* (Carbondale: Southern Illinois University Press, 1984), pp. 32–33.

41. For a discussion of the budget issues of the Trident program, see *The United Kingdom Trident Programme.*

42. *1985 French Defense Facts and Figures,* Service d'Information et de Relations Publiques des Armees, 1985.

PART FOUR

Soviet Strategic Policy and the Impact of SDI

8

The Impact of the SDI on Soviet National Security Policy

Daniel Gouré

As viewed by Moscow, the U.S. Strategic Defense Initiative SDI represents a political, scientific-technical, and military challenge to the Soviet Union. The foremost Soviet strategic concern is that the United States will be able to develop a combined offensive-defensive posture that would not simply deny Moscow a preemptive counterforce option, but, more importantly, would provide the United States with unilateral options for employing nuclear weapons for political purposes. Strategic defenses, Soviet commentators argue, make particular sense for the United States in the context of that country's efforts to develop limited nuclear options and to maintain a deterrence umbrella over its European and Asian allies. Coupled with the deployment of new theater-nuclear systems capable of reaching targets in the USSR, strategic defense is seen as offering the United States the prospect of fighting nuclear wars while escaping the consequences. An effective strategic U.S. defense would undercut the decades-long Soviet effort to create a strategic force posture capable of countervailing or paralyzing U.S. military power.

The U.S. commitment to a fifteen- to twenty-year development and exploitation program to create a space-based defense also represents a scientific-technological challenge to the USSR. The ability to grasp and exploit the ongoing scientific-technological revolution, Soviet leaders assert, is a central, increasingly important factor in judging the relative balance of Soviet and U.S. military potentials as well as forces-in-being. The U.S. program poses a challenge to the commitment by the Communist party of the Soviet Union to ensure the continual modernization and maximum effectiveness of the Soviet Armed Forces. In order to maintain equal security and improve the strategic balance, the Soviet Union must be able to match and even exceed U.S. military technologies. Ensuring scientific-technological superiority is also an important element in the current competition between the two opposing social systems. Although the Soviet Union need not match the West in every respect, a demonstration of the ability to match or exceed the West in revolutionary technologies and to apply those advances to outstanding military problems is an important indicator of Soviet scientific-technical status.

Soviet interest in strategic defenses is also a reflection of the belief that offense and defense exist in a dialectical union. Soviet strategic planners do not absolutize any particular technology, either offensive or defensive. They argue that every improvement in offense technology eventually generates a corresponding change in the character and capabilities of the defense. Although nuclear weapons have come to dominate the modern battlefield, Soviet planners see a possibility that further progress in exploiting the ongoing scientific-technological revolution may result in a return to a more balanced offensive-defensive environment.

The Soviet Union's current massive strategic defense program underscores Moscow's continuing commitment to defenses even at a time when Soviet offensive capability is expanding both quantitatively and qualitatively. The current upgrade program is intended to exploit advances in defensive technologies and to fill existing gaps in the Soviet defense structure. In particular, Moscow has focused on closing the gap in low-level air defense, establishing a high-confidence early warning and tracking network, and exploiting improvements in ballistic missile defense (BMD) technologies. In addition, the Soviet Union has underway R&D programs in advanced technologies that include directed-energy weapons for air, BMD, and antisatellite defense. The current Soviet program could permit the deployment of multilayered air and missile defenses starting in the later 1980s.

Moscow's reaction to the U.S. SDI suggests that despite ongoing criticisms of the feasibility and cost of such defenses by Soviet commentators, Moscow is concerned that the system might work, at least insofar as it would alter the current strategic balance and pose a major problem for Soviet strategic planning. If the Soviets were convinced that the system could be readily negated, or would prove impossible to implement, they would welcome the diversion of enormous amounts of money and resources to an unworkable system. However, even if the space-based element of SDI were unsuccessful, there remains the possibility of deployment of other types of strategic defense. Moreover, development of a defense against ballistic missiles may increase the likelihood that the United States also would seek to develop air defense and even a nationwide civil defense program. The cost to the USSR of countering or defeating U.S. strategic defenses undoubtedly would be quite high, even if it were only a fraction of the cost of the defense itself. Such resources are increasingly scarce given the Soviet Armed Forces' commitment to modernizing its conventional and naval forces.

Moscow clearly would prefer to stave off an intense U.S. effort to achieve enhanced strategic defenses. For this reason, the Soviets are unlikely to initiate modification or abrogation of the Antiballistic Missile (ABM) Treaty. Instead, Moscow would prefer to let the United States make the first move, thereby retaining the option of agreeing to amending the treaty but at the right "price." Current Soviet strategic defense efforts provide the basis of deployments of ABM and antitactical ballistic missile (ATBM) systems in the future, should this become necessary. Rather than an overt defensive arms race, Moscow would prefer the current limited competition, which provides room for "ambiguous" deployments such as the new phased-array radar in Siberia. Moscow would prefer to hold off a competition in defensive technologies until at least the 1990s when its current modernization effort would be completed and ongoing technology programs would be more developed.

In the event the United States does proceed with some form of space-based defense, Moscow would respond in a number of ways. Offensively, Moscow is likely to accelerate its current force modernization efforts, including mobile intercontinental ballistic missiles (ICBMs), longer-range submarine-launched ballistic missiles (SLBMs), strategic bombers, and cruise missiles. In addition, the Soviets would intensify efforts to develop antisatellite weapons and other countermeasures to projected U.S. defenses. Although Moscow has stated publicly that it believes countering proposed SDI technologies, is accomplished easily, how well such countermeasures will work in practice is far from clear. Developing and deploying numbers of complex countermeasures will be costly to the Soviet economy, will inject (undesirable) complexity into Soviet strategic planning, and may be unsuccessful.

Soviet Military Strategy, Technology, and Force Posture Requirements

It is generally recognized in the West that for all their apparent similarities, U.S. and Soviet strategic policies are fundamentally different in a number of critical respects.[1] One of the most important of these differences is Moscow's seeming rejection of Western notions of stable deterrence. This is not to say that deterrence of nuclear war is not a basic objective of Soviet military strategy and force development. Rather, there exist conflicting national views regarding what constitutes a stable balance and the role of the two sides in ensuring stability while maintaining their own security. Acceptance of the Western notion of mutual assured destruction (MAD), the Soviets argue, would ultimately place Soviet security at risk by an overreliance on U.S. "rationality."[2] Moreover, what deters best, according to Moscow, is not the ability to retaliate, but the capacity to deny an opponent its strategy by defeating its forces.[3] Because capitalism is the source of all wars and the possibility of war exists so long as capitalism and socialism coexist, Moscow must prepare for the possibility of war. Stability, from Moscow's point of view, depends on the ability of the Soviet Union to deter the West from aggression by continually enhanced offensive and defensive military might.

Soviet interest in developing capabilities to successfully prosecute a nuclear war is a logical extension of this notion of deterrence. The capabilities required to defeat an aggressor armed with nuclear weapons and ballistic missiles constitute a formidable deterrent. However, such capabilities permit the USSR to entertain the idea that it in fact might "win" such a war insofar as the Soviet Armed Forces were able to deny the opponent its objectives while simultaneously destroying that opponent's war-waging capabilities. The employment of offensive, even preemptive, actions against an aggressor thus is ideologically justifiable and, particularly in the nuclear age, strategically necessary. A Soviet military theoretician stated that "The founders of Marxism never linked the concept of 'defensive war' and 'defensive operations.' They understood that a state which is conducting a defensive war, that is, just war, has a legal right to wage offensive operations, as dictated by the course of military events."[4]

Given the inherent destructiveness of nuclear weapons, the most effective means of denying an opponent a high-confidence first strike and limiting the potential damage to the Soviet Union is by the preemptive destruction of those forces capable of attacking the USSR. The Soviets long have been interested in acquiring the capability for counterforce targeting.[5] With the deployment of land- and sea-based ballistic missiles by the United States, the requirement for preemptive was added to Soviet doctrinal expectation. Whereas the United States has argued that a preemptive counterforce capability was inherently destabilizing, the Soviets felt that such a capability, although technically difficult to achieve, nonetheless was strategically desirable. Although Soviet discussions of the role of preemption generally have been veiled, acknowledgment of preemption as a defense planning objective nevertheless can be found. In a recent monograph, Marshal N. V. Ogarkov stated that

> an important element of the military-technical basis of Soviet military doctrine, under conditions of the intense development of rocket-nuclear weapons and the possibility of their surprise employment by the enemy,

> is the necessity of maintaining the Armed Forces of the USSR in the highest combat readiness, insuring the timeliness of its actions for the repulse of the surprise attack by the opponent, the carrying out by it of powerful answering strikes and the successful fulfillment of immediate problems in the defense of the socialist Fatherland. The question is to possess not merely defenses, countervailing the aggressor by corresponding passive means and methods of defense, but to inflict on him crippling answering blows and to destroy the enemy under all conditions established by circumstances.[6]

Recognition of the potential destructiveness of nuclear war and of persistent uncertainties about the possibilities for successful preemption have given added impetus to the long-standing Soviet interest in strategic defense and damage limitations.[7] In the initial postwar period, strategic defenses, primarily air defense, served as a partial means of negating the effectiveness of the U.S. strategic threat. During the past four decades, the Soviet Union has devoted considerable attention to the development of a system of national defense involving antispace defense, BMD, air defense, and civil defense. Strategic defenses, the Soviets argue, serve to buttress the credibility of their strategic deterrent. Moreover, strategic defenses also enhance the political utility of Soviet military power. The ability to threaten military power in response to hostile actions depends, in part, on a demonstrated capability of the state and its population to deal with possible nuclear strikes. Strategic defense is described in Soviet writings as "one of the decisive strategic factors ensuring the ability of the state to function in wartime and, in the final analysis, the attainment of victory."[8]

The intensification of the scientific-technical competition between East and West seemingly has reinforced the Soviet belief that strategic defenses inevitably will be developed. Some sources even have noted that the current phase in the revolution in military affairs has reached the point of its own negation in which systems based on current technologies only can be quantitatively increased. According to one Soviet observer, "The time has come when existing weapons systems, one can say, have exhausted themselves, and are only being quantitatively increased. At the same time, the possibility of creating new weapons systems based on new principles has arisen."[9]

The process, Soviet military analysts acknowledge, is not necessarily clean and simple—new systems arise that absolutely and totally negate the effectiveness of older ones. The appearance of the negating system usually results in changes to the existing weapons and forms of warfare that permit their continued effectiveness, albeit in an altered manner. Thus, the development of firearms led to the demise of armored knights but saw the development of light cavalry. With respect to nuclear weapons and the defenses against them, Soviet writings seem to predict that the latter will inevitably be developed and will cause changes in offensive systems deployment.

> The experience of past wars convincingly demonstrates that the appearance of new offensive weapons has always inevitably led to the creation of means to counter them. . . . This applies fully also to nuclear missile weapons, whose creation and rapid development forced military-scientific thoughts and practice to actively develop means of countering them. In their turn, the appearance of means of defense against weapons of mass destruction caused improvements in means of nuclear-missile attack. All this confirms the conclusion that the continuous struggle between means of attack and defense, i.e., of weapons and equipment, is one of the main sources of development of military affairs as a whole.[10]

The absence of antiballistic missile defenses did not result in a diminution of attention or resources to other areas of strategic defense. Continuing attention to strategic defense of all kinds is justified in Soviet writings as a reasonable response to the potentially catastrophic consequences of a nuclear war.[11] The continuing heavy investment in air defenses was justified on the grounds of the large fraction of U.S. retaliatory power deployed on bombers and the presence of hundreds of NATO and U.S. forward-based aircraft. A careful reading of statements regarding the importance and effectiveness of strategic defenses suggests that the Soviets recognized that these defenses possessed significant weaknesses or gaps and that under present conditions they could expect to attrite only a relatively small fraction of attacking weapons.[12] Moreover, Soviet defense strategy relied for its effectiveness on the operation of offensive systems as well as of layered defenses. In the absence of successful preemption and the resultant weakening of the retaliatory strikes, it does not appear as if Moscow were extremely confident in its ability to survive. Thus, in one civil defense manual, the author notes that

> while the Armed Forces will insure the defense of the homeland with active means—by destroying the attacking weapons themselves (missiles at the launch sites, aircraft on airfields, submarines at sea) or by intercepting the means of destruction on their course to the targets, civil defense, by carrying out defensive measures to protect the rear areas, must achieve maximum weakening of the destructive effects of modern weapons on it.[13]

Recent Soviet statements also suggest that Moscow increasingly is doubtful about its ability to achieve preemption on a scale that would provide sufficient military benefits to end the war effectively. The limited effectiveness of a strategic preemptive counterforce strategy, the creation of new forms of warfare, and the expansion of zones of possible operations all are identified as new elements in the strategic equation that is challenging Soviet strategic planning. In an interview in *Krasnaya Zvezda,* published on May 9, 1984, Marshal Ogarkov noted that the proliferation of nuclear weapons had created an interesting strategic paradox.

> On the one hand, it would seem, a process of steadily increasing potential for the nuclear powers to destroy the enemy is taking place, while on the other hand there is an equally steady, and I would say, even steeper reduction in the potential of an aggressor to inflict a so-called "disarming strike" on his main enemy. The point is, with the quantity and diversity of nuclear missiles already achieved, it becomes impossible to destroy the enemy's systems with a single strike. A crushing retaliatory blow against the aggressor, even by the limited quantity of nuclear warheads remaining to the defender—a strike inflicting unacceptable damage—becomes inevitable under present conditions.[14]

Marshal S. Akhromeyev, in the Communist party journal *Kommunist,* took this point one step further. He argued that the potential consequences of a global nuclear war were becoming even more dangerous while simultaneously claiming that a first strike could not disarm the target country. He also showed surprising caution in describing the possible extent of the war, targets, and its consequences.

> The opposing sides now have accumulated such a quantity of nuclear weapons that their use can lead to catastrophic and irremediable consequences. If the imperialists succeed in unleashing a world war, it would be of global proportions for mankind, would extend over the entire territory of the world and could be a war of extermination (large cities and many economic installations could be destroyed) and the population and groups of armed forces would suffer enormous losses.
>
> But at the same time, the aggressor is no longer able to inflict a disarming strike and act with impunity.[15]

Despite strong evidence that the Soviet military eschews initiation of central strategic war as a desirable mechanism for achieving Soviet aims, Moscow has not ceased planning for such a contingency or the best means of employing such weapons to achieve its objectives if such a war should occur. Commentaries by Ogarkov, Akhromeyev, and others appear to focus primarily on two contingencies; a theater conflict and a general strategic/global war. Although Moscow would clearly prefer a controlled conflict, it cannot be guaranteed such an outcome, particularly in these two situations. Moreover, the Soviets continually stress the possibility that the United States will attempt to resolve outstanding political and military disputes with the socialist community by resort to a nuclear war, including a first strike. If history teaches the Soviet leadership any single lesson, it is that things rarely go as planned. Hence, although Soviet military planning admits the possibility of a controlled theater conflict, in the event of a general outbreak of hostilities with the West Moscow still expects the war to become nuclear. Moscow still requires the Soviet Armed Forces to achieve their primary military missions: destruction of the enemy's ability to wage war and protection of the Soviet homeland. This, in turn, means a continuing emphasis on strategic offensive and defensive forces.

It might be argued that the evolving Soviet strategic view also should increase interest in the development and deployment of strategic defenses, including forms of ballistic missile defense. Soviet writings relate the problem of the growing complexity of military operations globally and within a theater with a requirement to enhance the synergy between force elements. These writings suggest an evolution in Soviet strategic thinking away from an emphasis on a single massive nuclear strike and toward a series of strategic operations involving various branches of the armed services, either alone or in combination. The strategic offensive constitutes but one form of military action. Strategic defense includes the timely repulse of the enemy's offensive strategic groupings, the inflicting of heavy losses on enemy forces, the shielding of important regions, and the creation of conditions for transitioning to the strategic offensive. In both cases, success depends on the close cooperation of various types of combat forces including PVO, the navy, air force, ground forces, and the Strategic Rocket Forces.

Soviet interest in strategic defenses also is being fueled by a sense that the scientific-technical revolution may be on the verge of radically altering the ability to defend against long-range offensive systems. The deployment of new theater nuclear weapons, development of highly accurate conventional strike systems, and research involving "new physical principals" all are cited by Soviet writers as evidence of a potentially revolutionary change in future forms of combat. In a recent article, Marshal V. Kulikov, former commander in chief of the Warsaw Pact, noted that

> Accelerator, laser, and radio frequency weapons and high explosive ammunition are being created. Reconnaissance and strike systems are being equipped with highly effective intelligence systems, radio-electronic

systems, and high accuracy terminally-guided warheads. Matters are moving towards a situation where conventional weapons approach nuclear weapons in terms of effectiveness and uses. Space is being militarized.[16]

A similar expression of concern was forthcoming from Marshal Ogarkov: "The rapid development of science and technology in recent years create real preconditions for the emergence in the very near future of even more destructive and previously unknown types of weapons based on new physical principles."[17]

The importance of changes in weapons technologies as they affect strategic planning is central to what the Soviet's have termed "the revolution in the military affairs." This revolution, generally a function of the development of nuclear weapons, long-range delivery means, and modern command, control, and communications (C^3) systems, resulted in a qualitative change in military affairs and, simultaneously, in economic planning, foreign affairs, and scientific-technical policies. The continuation of this revolution—indeed, its broadening into a general "scientific and technological revolution"—is considered by Soviet leaders one of the general patterns of the postwar evolution in military art. Equally important for Soviet strategic planners is the seemingly quickening pace in the development of new weapons systems. Soviet military writers now claim that revolutionary changes in military technology take place every twelve to fifteen years.[18] In his 1982 assessment of the evolution of military affairs in the current era and of the ongoing revolution in military affairs, Marshal Ogarkov noted five key elements.

First, the scientific and technological revolution, which imposes heightened demands on the qualitative characteristics of military equipment and weapons and on the search for new modes and forms of conduct of combat operations, is exerting increasing influence on the evolution of military affairs.

Second, there is taking place an acceleration in the pace of development of military equipment and weapons, with a shortening of the time intervals between qualitative leaps forward in the development of various areas of military affairs, which in turn is affecting the pace of development of military affairs as a whole.

Third, there is taking place an increase in the importance of strategic weapons, which are today capable of exerting direct influence on the course and outcome of a war, and consequently the importance of operational-strategic command and control agencies is also increasing.

Fourth, troop control processes are becoming more complex, which requires a fundamentally new approach to the organization of structurally precise control systems and providing them with the requisite modern control hardware.

> And finally, aerospace, which gives modern operations a three-dimensional character and depth, is assuming an ever increasing role in combat actions and operations.[19]

The Soviets argue that progress in the development of military technologies is inevitable. Moreover, the Soviet Union as a matter of state policy is committed to ensuring against technological surprise or an unfavorable shift in the strategic balance. Such a commitment is accomplished by applying the fruits of national scientific-technical achievements and economic power to the enhancement of the capabilities of the Soviet Armed Forces. Increasingly, according to Soviet sources, the competition between the two opposing military blocs is shifting from one that is quantitative in nature to one that is qualitative.[20]

High-level Soviet statements suggest that Moscow also would respond to any U.S. strategic deployments, including that of defensive systems, particularly if it perceived those deployments as threatening the current strategic balance. While renouncing the idea of military superiority, former Soviet defense minister Marshal D. F. Ustinov stated that "neither will we allow anyone to upset the military-strategic equilibrium that has been achieved."[21] Similarly, in commenting on the U.S. strategic modernization program, the chief of the General Staff, Marshal Akhromeyev, declared that the USSR will not allow the United States to attain military superiority over it, and that "as long as the aggressiveness of imperialism is growing, the danger of a new world war breaking out is also increasing. Therefore, following the Leninist peace-loving foreign policy course and persistently struggling to curb the forces of war, our party and the Soviet government continue to devote unremitting attention to strengthening the country's defense capability in every way possible."[22]

Particularly threatening from Moscow's point of view is the combination of U.S. strategic defenses, primarily those capable of providing countervalue protection, and the string of new offensive nuclear forces including the Pershing II, ground-, sea-, and air-launched cruise missiles, B-1, Trident D-5, and the MX intercontinental ballistic missile (ICBM). Soviet observers argued that strategic defenses make the most sense in the context of a first-strike counterforce capability.[23] The present minister of defense, Marshal of the Soviet Union S. Sokolov, in an interview for the Soviet news agency TASS, argued that the U.S. objective was to overturn the existing military-strategic balance.

> For that purpose, efforts are underway to improve the accuracy of U.S. nuclear systems capable of hitting our retaliatory strike forces, primarily intercontinental ballistic missile (ICBM) silo launchers. Conditions are being created for a sudden nuclear attack with the help of Pershing 2

> missiles being sited in Western European countries and long-range cruise missiles of different basing modes. . . . Different means are being used to camouflage U.S. missiles and bombers in flight to limit to the maximum the possibility of their detection.
>
> The Pentagon is developing at accelerated pace new strategic offensive systems—two types of ICBMs, submarine-launched ballistic missiles (SLBMs), and two types of heavy bombers. Secretary of Defense C. Weinberger does not bother to conceal that, while developing a space anti-missile system, the United States will also have a powerful strategic triad: for dealing a devastating nuclear strike that threatens the existence of the Soviet Union.[24]

A well-known Soviet analyst with strong ties to the military described the Soviet image of the new U.S. strategy as follows.

> The doctrinal position of the United States is not merely a desire to guarantee first (disarming) strike potential. The new American strategy is primarily intended to guarantee so-called escalation dominance: to give the United States the ability to force its own opponent in a nuclear missile exchange to accept the U.S. "rule of exchange" (or, as the Americans put it, "rules of the game"). American strategists realize that the other side will always, even under the worst circumstances, be able to deliver a strike against the urban and industrial complex in the United States. This is why the American "proposal" to spare the cities and deliver strikes only against military facilities clothes an extremely pragmatic consideration with the external trappings of humanitarianism: this would force the other side to accept a form of nuclear "duel" in which the United States' partial advantages (in the form of some kind of qualitative advances) would become the deciding factor, giving the United States an opportunity to disarm the other side and take away its potential for effective deterrence.[25]

According to the Soviets, Washington intends to expand on this already awesome capability by attempting to deploy strategic defenses. This view often takes on an air of inevitability to Soviet military theoreticians because, after all, "if potential opponents possess weapons of mutual destruction, decisive advantage goes to the side which manages to create a defense against it."[26] This view is reinforced by a growing recognition in the Soviet Union that the nature of modern nuclear forces, particularly their various deployment modes, hardness, and redundancy, means that no first strike can hope to prevent some retaliatory strikes by the opponent.[27] The search for a first-strike strategy therefore logically also must include an effort to develop defenses with which to attenuate the inevitable retaliation.

Despite Soviet statements in recent years to the effect that there can be no victory in a nuclear war and no hope of escaping retaliation, the political-perceptual impact of a U.S. effort to deploy new military capabilities, including strategic defenses, is likely to be assessed by the Soviet Politburo and General Staff as sufficiently threatening to Moscow's objectives so as to warrant countervailing measures.

Soviet Responses to the U.S. Strategic Defense Initiative

U.S. efforts to develop strategic defenses must plan against current as well as future threats. The problem of threat assessment is all the more difficult because the USSR will undoubtedly react to U.S. efforts with R&D and force acquisition initiatives of its own, which could substantially alter the threat against which strategic defenses must operate. Strategic defense and offense countermeasures planning constitute a strange pas de deux in which neither side wishes to lead but instead to follow. The inchoate nature of current U.S. strategic defense plans, their dependence on the ability to resolve a host of outstanding technical and operational issues, and the likelihood that changes in Soviet capabilities may produce changes in U.S. strategic defenses efforts serve to increase Soviet difficulties in formulation of their threat estimates, R&D programs, and acquisition strategies. As the U.S. program becomes better defined and as deployments are planned, the Soviets will be able to identify more readily potentially effective countermeasures.

Until recently, the Soviet Union had reason to view the current strategic and theater balances as favorable to the pursuit of Soviet objectives. The attainment of nuclear parity, according to Moscow, had constrained severely the ability of the United States to threaten the Soviet Union or to enforce U.S. global objectives by military means.[28] For this reason, war between the two systems was viewed as unlikely. The sense of U.S. homeland vulnerability created and reinforced by Soviet strategic forces produced a tendency toward realism in U.S. dealing with the USSR. This tendency was exaggerated by U.S. experiences in Vietnam and Iran.[29] As defined recently by the Soviet minister of defense Marshal Sokolov, existing strategic and theater balances are important elements of Soviet national security and foreign policy: "The existing military-strategic balance between the USSR and the USA, between the Warsaw Pact and NATO is a historical achievement of the socialist community, and indispensible prerequisite of its security. . . . The existence of this balance deters the USA's imperial ambitions and prevents it from achieving world dominance."[30]

The U.S. Strategic Defense Initiative is considered one component of a long-standing U.S. effort to develop a nuclear war-waging capability. Soviet observers see the SDI, coupled with development of new, more capable offensive systems, as providing the United States with the means, in theory, of conducting a successful nuclear offensive while protecting the United States and its allies from inevitable retaliation. One Soviet observer described the Soviet view of the relationship between the SDI and offensive nuclear force modernization as follows:

> Plans are being made to attempt, by the first use of nuclear weapons, to reduce as far as possible the USSR's defense capability and its nuclear missile defense first and foremost—and to present the Soviet Union with the dilemma of either agreeing to expand the scope of nuclear war and its unpredictably destructive consequences, or capitulate. . . .
>
> An even more sinister option is being contemplated for the more distant future: to disable the USSR through the joint effect of strategic offensive and defensive capabilities. This is the purpose of the space-based ABM defense system which is now being developed.
>
> The practical implementation of this option could, or so it is claimed, not only result in the destruction of the USSR's Armed Forces, its key industrial targets, its centers of state and military control, but also protect the United States against the results of a nuclear catastrophe.[31]

The dominant Soviet concern is clearly the deployment of weapons in space. Although Moscow has been highly critical of any U.S. effort to deploy new strategic defenses, the intensity of Soviet reactions to the prospect of national deployment of weapons in space suggests that the Soviets believe space to be a unique vantage point for strategic operations. Likewise, they believe that space may constitute the most advantageous position from which to strike targets on the ground and that the U.S. SDI program also will involve development of both nuclear and exotic weapons for this mission.[32] One Soviet commentator, in a statement typical of this type, declared that the United States had under development weapon systems to hit targets on the ground: "As became known recently, there is a secret Pentagon program entitled Prospective Strategic Missile Systems. In the framework or this program, high-velocity shells are being developed to be fired from space-based electromagnetic guns and aimed at ground targets with optical lasers."[33]

Soviet commentaries on the SDI suggest some uncertainty within the USSR as to the seriousness of the U.S. commitment and the best means of responding in the event the United States does deploy some form of BMD. Even if the United States does proceed with R&D on advanced technology defenses, Moscow is aware that the likely deployment horizon for such a system would be well into the 1990s, at the earliest. A change in administration could reverse the current commitment of strategic defense. Under such conditions, an early Soviet response would not be politically advisable. However, at least one source predicted the first deployment of U.S. directed-energy weapons by the early 1990s.[34]

Another explanation for the Soviet reaction may be that the Soviet military was not particularly surprised by the U.S. decision regarding SDI research and development and may feel that it already has programs underway with which to offset or match a U.S. effort. Senior Soviet military officers long have argued that there is an inherent dialectical relationship between offensive and defensive technologies and that, in the words of Marshal Ogarkov, this is "one of the principal sources of development of military affairs as a whole."[35] Soviet military technical analysts have noted that the United States has been working on advanced weapons systems for a long time.[36] Additionally, some Soviet officers, such as the former chief of the Strategic Rocket Forces General Tolubko, appear to believe that the USSR has matched—and can continue to do so—U.S. offensive and defensive weapons developments.[37] Both Akhromeyev and Sokolov have asserted a requirement capability to negate U.S. defenses.[38]

Another, more revealing but pessimistic statement was made by a senior Hungarian army officer. In a national theoretical conference called to analyze the imperialist threat in the 1980s, Lieutenant-General Ferene Szues Desenihed warned that the U.S. strategic defense initiatives must be taken very seriously.

> The question of space armaments must be taken seriously, even if some elements of it belong only to the realm of science fiction stories, and if for the time being they exist only in laboratories and on the desks of scientists. It must be taken seriously because to an increasing extent these scientific fantasies can be realized. *It is only a question of time and money* (emphasis added).[39]

The U.S. debate regarding criteria for deployment of strategic defenses, deployment timing, and the nature of the potential transition is viewed with evident suspicion by Soviet observers. The Soviets apparently consider this a debate about deployment strategy, not a debate about the feasibility or advisability of deploying the SDI. There is apparently some belief that the United States first might deploy even an only partially effective (or relatively ineffective) defense. "One advocate of 'Star Wars' gives assurances that the system will be deployed only when it is 'invulnerable,' while the other claims that it will be deployed in any case, while constantly expanding and improving the invulnerability of its components."[40]

Initial Soviet concern about weapons in space was tied to the growing U.S. interest in developing its antisatellite (ASAT) weapons capability.

Moscow considers the U.S. ASAT program one aspect of the U.S. effort to achieve space dominance. The Soviets consistently have criticized the current U.S. ASAT program as an attempt to create a strategic advantage that would lead to a new round in the arms race. Soviet commentaries focus less on any specific target for the U.S. ASAT than on the theme that attacks on military space systems are an important, perhaps even necessary, element of a first-strike strategy. The Soviets suggest that the U.S. ASAT program is part of the overall U.S. effort to create a strategic situation in which Washington would be able to conduct a limited nuclear war without the risk of direct retaliation on U.S. territory. One commentator noted that "it would appear that thoughts of creating a space weapon are unavoidably linked to the use of nuclear weapons and other types of weapons of mass destruction—otherwise, the expensive space technology would be equivalent to using a cannon to shoot at sparrows."[41] Another commentator noted that the current U.S. military space program was the result of a planned and logical progression from basic launchers to reconnaissance satellites to combat support systems and now to space weapons.[42]

Among Soviet concerns regarding ASAT weapons is the potential threat posed to surveillance satellites in general and, specifically, to national technical means associated with the verification of arms control agreements. In addition, the Soviets have expressed concerns, which echo prior Western criticisms, that efforts to impede or destroy surveillance satellites "might result in a situation where each state, deprived of trustworthy data about the opposing block's military preparations, would proceed to develop its armed forces from the 'worst-case' variant in its estimations and overreact to any, if unchecked and unlikely, data."[43] In general, Soviet critiques of the U.S. ASAT system suggest that the Soviets expect its use to be against near-earth orbit systems. However, the Soviets worry that current efforts will be expanded to include systems capable of attacking satellites at higher orbits, including geosynchronous orbits.

Soviet analyses, both before and after President Reagan's March 1983 speech, generally have adopted the line that the United States has made a basic decision to achieve a damage limitation capability based on a combination of offensive and defensive capabilities.[44] The Soviets have criticized the SDI on a number of grounds. It would destabilize the strategic balance, make a nuclear war more likely, create a new round in the arms race, and create incentives for the side possessing such a system to initiate a first strike. Some Soviet critics point out that no strategic defense can be expected to operate with 100 percent effectiveness. Hence, the optimum use of such a system is in conjunction with a counterforce first strike that limits the number of second-strike weapons the defense must deal with.[45] It is accepted by all Soviet commentators that the SDI program will be directed toward creation of a national damage-limiting capability.

Soviet commentaries on the U.S. strategic defense program have tended to differentiate defense of point targets, particularly missile silos, from a nationwide defense usually described as space based.[46] Neither is viewed as stabilizing. Indeed, one Soviet commentator went so far as to declare that the effort to protect U.S. missile silos by any means is destabilizing and a violation of the principle of equal security.

> And now the Pentagon intends to revive its "deterrent capability" by shielding American missile launch silos with space-based weapons. But no matter how many the reassurances offered by Star Wars supporters that this is a purely defensive system, one finds it impossible to believe them. *Any reduction in the vulnerability of strategic weapons designed to destroy enemy launching sites is tantamount to disturbing the equilibrium achieved, to creation of a new first-strike capability* [emphasis added].[47]

In discussing Soviet responses, Soviet scientists suggest three classes: expanded offense, offensive countermeasures, and a restructuring of Soviet offensive forces. A May 1984 study notes the possibilities of ground-based ballistic and laser-driven, antisatellite weapons, "clouds" of obstacles (heavy balls), false launches to force the system to engage early, and space mines.[48] The latter, in particular, appears to be drawn from well-known Western sources on potential Soviet countermeasures rather than from any Soviet studies. Another Soviet source suggests that the creation of a space-based, antimissile system implies that "means of breaking through it will be improved at an accelerated pace—ICBM, submarine ballistic missiles, etc."[49] It is interesting that this source also suggests that a quantitative buildup would occur, but not in ballistic missiles.

> Creating a large-scale anti-ballistic missile system can provide a strong stimulus for a quantitative buildup of strategic delivery means and nuclear warheads—notably, the strategic cruise missiles, including the sea- and land-based ones, the deployment of which it is exceedingly difficult (if not at all impossible) to control with natural technical means of verification.[50]

More recently, some Soviet commentators have begun to identify the specific vulnerabilities of strategic defense concepts and to suggest possible Soviet countermeasures. The ubiquitous academician Y. Velikhov noted in one anti-SDI commentary that "the individual components of the space ABM system are extremely vulnerable since they have parts that can be easily destroyed, thus crippling the entire system."[51] In another article he suggested that a constellation of boost-phase defenses may be vulnerable to a combination of defense suppression and special launch tactics. "One need not destroy the whole of it. It is enough and easy to destroy just a small part of the system to make a hole." Velikhov went on to point out the possibility of saturating midcourse defenses as well: "There is yet another circumstance which makes this system vulnerable. Maybe nonspecialists do not know that it is not only warheads that can be launched to space with a missile. There will be tens of thousands of decoys which will attract a considerable portion of the said system."[52]

Georgy Arbatov, in an interview for a Japanese newspaper, argued that proliferation of offensive forces could overcome a defense or at least make it cost ineffective.

> The U.S. purpose is to nullify the Soviet Union's military efforts thus far and ruin it. However, missiles can be produced at a lower military expenditure than is required for SDI. Therefore, the Soviet Union does not need to imitate the United States. When the other side builds a defensive system, all you need is to have many more offensive means to make the system impotent.[53]

Soviet military leaders are more reluctant to specify courses of action in response to the SDI. In a recent interview in *Krasnaya Zvezda,* Minister of Defense Marshal Sokolov, in response to questions on the topic of countermeasures to the SDI, was noticeably cryptic.

> If the United States commences the militarization of space and thereby undermines the existing military-strategic equilibrium, the Soviet Union will be left with no other choice but to adopt countermeasures to restore the situation.
>
> This could be measured in the sphere of both defensive and offensive arms. It goes without saying that the USSR will choose the means of action which most accord with the interests of its defense capability, and not those towards which figures of Washington would like to incline it.[54]

The character of the Soviet response to the U.S. ASAT and SDI programs suggests that although the USSR does not perceive a specific threat to Soviet satellites or strategic offensive forces at present, it is extremely concerned about the uncertainties created for Soviet strategic and technological planning and the possibilities that the United States might achieve a technological breakthrough that would negate the utility of the Soviet nuclear arsenal. The degree of concern in Moscow can be judged by the extent to which previous issues, such as ground-based ABM, Long-range Intermediate Nuclear Forces (LRINF), and even ASAT, have receded into the background in Soviet policy statements and political commentaries. However, Soviet analysts do recognize that there is an extremely long lead time associated with the deployment of a workable space-based strategic defense. According to one source citing unknown experts, "The first generation of space and ground-based lasers, particle beams, and high-power microwave weapons is likely to appear in the late nineties or at the turn of the century."[55] During this period, the Soviets have a number of options open to them: a space treaty, buildup of offensive nuclear forces and changes to the force mix, development and deployment of countermeasures, and creation of a Soviet space-based weapons capability.

Marshal Akhromeyev suggested in a recent article that because of the relationship between U.S. strategic defense and Soviet security requirements, an offensive response by the USSR to deployment of such defenses was all but inevitable. Although phrasing his argument in defensive terms, he made clear that Moscow's goal is to prevent the United States from altering the current, and from Moscow's view favorable, balance of power.

> The creation and deployment in space of strike arms will inevitably lead to an increase in the quantity of and to the qualitative improvement of strategic nuclear arms. A close interconnection objectively exists between offensive and defensive strategic systems. That is the logic of nuclear confrontation. The creation of the large-scale space ABM system contemplated in the United States has a clear aggressive point: This system is a most important element in the integrated offensive potential of the side that has created it, undermines strategic equilibrium, and provides the opportunity for the United States to deliver a first strike in the hope that the retaliatory strike against U.S. territory can be averted.[56]

Most identified countermeasures to strategic defenses would require costly, complex, large-scale alterations to the current and projected Soviet force posture. The simplest solution, depending on the character of the defense, would be proliferation of additional offensive weapons, particularly ballistic missiles and reentry vehicles (RVs). This might enable the offense to overwhelm and saturate the defense. However, against a layered defense, particularly one employing boost-phase weapons, proliferation alone is unlikely to provide sufficient confidence of penetration of the defense.

In the event defenses are deployed in space, the Soviets could attempt to destroy or degrade their functions. Even supporters of the concept of strategic defense have noted the potential vulnerability of platforms in space.[57] A principal countermeasure to space-based defenses is likely to be antisatellite weapons. The Soviet ASAT program has moved at a desultory pace. Although based on relatively old and reliable technologies, the system may be undergoing modification in the form of new orbital intercept maneuvering capability, better kill mechanisms, and a new targeting system (possibly based on infrared or electro-optical sensors).[58] The replacement of older radar guidance systems with modern sensors could indicate a need to differentiate between different objects in close proximity, such as a U.S. shuttle and a Soviet satellite. Although the present Soviet ASAT is capable of threatening satellites only in low-earth orbit, development of a heavy lift booster could permit attacks on satellites up to geosynchronous orbit. Recently, a senior Soviet military spokesman claimed that Moscow had developed a new ground-based direct-ascent ASAT.[59]

Many options are available, in theory, to perform the defense-suppression function: conventional and nuclear ASAT, ground-based lasers, space-based conventional, nuclear, and directed-energy weapons, and even "space debris" launched into the path of the space platform. The choice of suppressing agents depends on the character of the defense, the timing for such attacks, and the cost to the attacker. No single suppressant is equally effective against every architecture. It is likely that the Soviets would deploy a mix of suppressing mechanisms. It is equally likely that the defense would deploy a complex system, involving several sensor and kill mechanisms, each set in a unique configuration or constellation, thereby complicating the attacker's problem. The critical trade-off for the Soviets is between defense suppression and additional forces, including missiles, RVs, and decoys.

Since any antiballistic missile weapon possesses some inherent capability to perform an ASAT function, deployment of a Soviet SDI may be a particular threatening problem for the United States. The most dangerous situation may arise from a Soviet system employing space-based, directed-energy weapons. The long-range, rapid retarget time and potentially high lethality of a directed-energy ASAT/BMD weapon would make coexistence of two such defenses extremely difficult.

Deploying defense suppression systems will be costly, particularly because such systems do not contribute directly to the attainment of strategic objectives. Moreover, defense suppression adds complexity to Soviet attack planning by requiring an additional "precursor" phase to the actual launch of strategic forces. The Soviets also may have a problem in deploying defense-suppression countermeasures that support both a preemptive and a second-strike retaliatory strategy. Successful defense suppression probably requires precise timing and coordination with the operation of offensive forces. This may not be available if Soviet forces were required to launch on warning to strike second.

Most other proposed countermeasures exert virtual costs on the Soviet force posture. Measures to defeat boost-phase defenses, such as a fast-burn booster or booster hardening and coating, penalize Soviet throw weight. Penetration aids (penaids), such as decoys, may require substantial offload of RVs. In addition, the employment of penaids requires a prolongation of the postboost busing phase in which the bus itself is vulnerable to attack. The deployment of a fast-burn booster, new postboost vehicles, penetration aids, and the like would constitute a costly reconfiguration of the Soviet force posture.

The problem for the Soviets in devising countermeasures to strategic defenses is that they may be forced to spend extremely large amounts of money for no better end than to recapture current relatively unsatisfactory preemptive and warfighting capabilities. The vision of an extremely costly, futile arms race was identified as a major underlying concern in a recent Soviet commentary on the SDI.

> Transferring the arms race into space is dangerous because it constitutes enormous new multibillion dollar arms expenditures requiring enormous resources, thereby siphoning resources away from the real aims of those nations taking part in the armaments race. This is a very expensive affair. . . . This is an exhausting arms race, and this is why we have been systematically proposing to the United States that they give up this aim.[60]

Although the USSR may have some advantage over the United States in devising countermeasures to the SDI, this edge should not be overemphasized. For one thing, the USSR also must begin to make choices in the its R&D programs. This involves taking risks regarding Soviet assessments of the direction of U.S. programs, including the potential utility of countermeasures and the maturity of the required technologies. The Soviets would like to minimize these risks and control the timing of the arms acquisition process. Arms control is intended to serve this function. If the U.S. program proceeds rapidly, then the Soviets may be forced to take serious risks in their weapons acquisition policies. The USSR may deploy the wrong countermeasure, one unable to seriously challenge the defense and that can be countered by newer U.S. developments.

Although it is possible to argue that almost anything is possible for the Soviets during the next twenty or thirty years, this does not mean that they can develop SDI countermeasures that will be deployed at the right time and in the required numbers. Once the competition between offense and defense begins as a serious planning and national issue,

Moscow clearly expects this competition to be a continuing problem. Despite the contentions of some Soviet writers, no easy solutions are envisioned by military or technical leaders.

Soviet Strategic Defenses

Regardless of the future of the SDI, the Soviets can be expected to continue investing heavily in strategic defenses of their own. The Soviet Union possesses the largest, most complex, modern network of strategic defenses in the world. Although by no means capable of providing the USSR with a damage denial capability, this system is capable today of providing significant damage limitation against certain kinds of attack. The overall effectiveness of Soviet strategic defense must be judged in relationship to evolving Soviet and Western offensive forces and to likely exchange scenarios. Defenses of marginal quality are potentially effective when operated in conjunction with a counterforce first strike or against a limited nuclear attack.[61] Soviet strategic defenses also have exerted a heavy virtual cost from opposing offensive forces. Development of the B-1, air-launched cruise missile, Stealth technology, modernization of the B-52 fleet, the air-launched ASAT system, and longer-range SLBMs have been necessitated, at least in part, by improvements in Soviet strategic defenses. The geographic asymmetries between the United States and USSR also enable the Soviets to employ strategic defenses in support of the operation of theater forces or to buttress homeland defense by reallocation of regional capabilities.

Currently, the Soviet Union is in the process of significantly enhancing its strategic defenses. Particularly in strategic air defense, these improvements mark a significant step toward a national defense against low-flying penetrators, both strategic bombers and cruise missiles. The existing force of some 1,200 interceptor aircraft is being upgraded through a process of retiring obsolescing aircraft and replacing them with modern systems. The new MiG-29 Fulcrum, MiG-31 Foxhound, and SU-27 Flanker all are believed to have a look-down/shoot-down capability. Armed with modern effective air-to-air munitions and employed in connection with the new Il-76 Mainstay SuAWACS, these are a significant improvement over existing air defense systems.[62] In addition, Moscow is deploying widely the SA-10, a long-range, mobile, high-acceleration system and has begun deployment of an even more capable surface-to-air missile (SAM), the SA-12. Both systems have the capability to engage fast-moving, low-flying objects as well as to defend against objects in the medium-to-high altitude regions.[63] The effectiveness of such deployments can be imputed from the U.S. Air Force's decision to truncate the acquisition of first-generation, air-launched cruise missiles in favor of more rapid introduction of a "Stealthy" version.

Improvements to Soviet defenses against ballistic missiles have focused on three principal areas: (1) upgraded and expanded early warning, threat-assessment, battle-management surveillance systems and associated command, control, and communication (C^3); (2) modernization of the aging Galosh system around Moscow; (3) and development of components for a relatively rapidly deployable or even mobile ABM system. The current C^3 upgrade program involves space- and ground-based early warning systems, air defense, and BMD battle-management radars, and possibly, an improved capability for battle management for space warfare.[64] Enhanced early warning of ballistic missile attack is provided by the new Soviet geosynchronous early warning satellites, first deployed in 1975. This new system is a significant improvement over previous Molniya elliptical-orbit satellites. In addition, the Soviets have deployed some 7,000 air defense early warning and battle management radars. For ground-based warning and tracking of ballistic missiles, Moscow has deployed eleven Hen House early warning radars and two over-the-horizon (OTH) radars along the periphery of the USSR. In addition, the Soviets have built or are constructing at least six extremely large, phased-array radars. These new radars offer significantly improved target tracking and ensuring the complete coverage of the periphery of the USSR against attack from any direction.

Construction of the latest large phased-array radar in central Siberia has raised concerns regarding its function and legality under the 1972 ABM Treaty. The location of the new radar and its northeast orientation suggest that it is intended to perform an ABM battle management function, a violation of the restrictions in the ABM Treaty.[65] The U.S. government officially considers the radar a violation of the legal obligations under the ABM Treaty.[66] This radar would appear to be located so as to provide the first phase of a defense against a U.S. threat to Soviet ICBMs deployed in central Siberia. The Soviet response to U.S. queries has been to claim that the new radar is not for ABM battle management but for radar calibration of test launches from the test complexes at Tyuratam and Sary Shagan. Other sources have suggested that this radar is intended to provide tracking and targeting information for the Soviet orbital antisatellite system.[67] Although the new radar is of relatively little immediate military significance given the current state of Soviet ABM defenses, should Moscow decide to expand its current ABM deployments to a regional or nationwide basis, the existing constellation of ballistic missile tracking and battle management radars could provide substantial targeting assistance, particularly if that system were a simple, rapidly deployable ABM.

The long-watched improvements to the Moscow Galosh system are now nearing completion. Existing Try Add, Cat House, and Dog House radars are either being replaced or supplemented by new systems, including a phased-array ABM engagement control radar at Pushkino, north of Moscow.[68] In addition, the existing sixty-four ABM launchers (reduced to thirty-two during preparations for the transition to the new system) are being replaced with new interceptors. The new ABM system involves two different silo-based interceptors, the SH-04, a long-range exoatmospheric missile, and the SH-08, a high-velocity missile intended to provide endoatmospheric defense.[69] The Soviets may expand the upgraded program to the full 100 interceptor launchers permitted under the ABM Treaty. This system is remarkably similar to the proposed Safeguard ABM system the United States considered deploying in the early 1970s. One of the most important possible differences between the now-defunct U.S. system and the emergent Soviet one is that the latter has reportedly been tested with rapid reload equipment.[70] The new Moscow ABM is expected to be operational sometime in the late 1980s.

There are also reports that the Soviets have developed, and possibly are producing, components for a mobile, rapidly deployable ABM system. The Flat Twin and Pawn Shop radars combined with both SH-04 and SH-08 in above-ground launchers or with the new SA-12 air defense missile are alleged to form the basic components of such capability.[71] The deployment of new phased-array radars in an ABM mode should be particularly worrisome given the possibility of Soviet preparations for relatively rapid deployment of an ABM system. These radars would provide necessary assistance to less-capable engagement radars associated with individual ABM sites. The idea of a rapidly deployable ABM should not be overstated. Although the various elements of the suggested system are transportable, they require large moving vehicles and substantial site preparation before deployment. Hence, the United States would probably be able to detect preparations for such a deployment long before the system was in actual operation.

The Soviet decision to match U.S. deployments in space would depend on Moscow's assessment of its ability not only to solve complex technical problems associated with directed energy weapons and advanced sensors but also those engineering and operational problems of large-scale operations in space. Moscow certainly would increase its attention to near-term space projects, such as the Soviet space shuttle, heavy-lift booster, and permanently manned space station.

In the past decade, Moscow has made major improvements in its capacity to employ satellite systems to perform a host of military support functions, including communications, surveillance, targeting, navigation, early warning, and meteorology.[72] In addition, the Soviet Union has made major strides toward its goal of establishing a permanently operating space station. A permanent station in space could be used for military purposes, including as an antisatellite battle station. At least two of the Soviet Salyut space stations are believed to have served as military reconnaissance platforms.[73] In support of its various military space programs, the USSR has been developing heavy-lift launch vehicles and may have developed two different types of reusable shuttlelike craft.[74]

An additional area of Soviet effort is likely to be enhanced C^3 for both air and antimissile defenses. The recent reorganization of Soviet strategic defenses in which the Air Defenses of the Ground Forces (PVO Sukhoputnikh Voisk) and the National Air Defenses (PVO Strany) were combined into a single entity, the National Air Defense Forces (Voiska PVO), will permit more rapid coordination of all air defense elements and provide an organizational basis for shifting resources more effectively to meet emerging threats. Such a move also could serve to enhance Soviet defenses against European-based and ballistic missiles that must overfly Warsaw Pact territory in order to reach Soviet targets.

There are reports that the Soviets also are pursuing development of a number of directed-energy weapons as part of their air, ballistic missile, and antisatellite defenses.[75] The possibility of employing directed-energy weapons in defense against threats in the air or in space is hardly new in Soviet thinking. Major-General I. J. Anureyev predicted in 1971 that the long-range development of strategic defenses would include development of both ground- and space-based directed-energy weapons.[76] It is likely that Soviet R&D in directed-energy weapons began no later than the early 1960s at a time when the United States was conducting research under the Project Defender, Saint, and Bambi programs. Some U.S. sources estimate that a Soviet ground-based, directed-energy weapon could become operational by the end of the decade and that a space-based system is possible by the year 2000. Soviet deployment of directed-energy weapons is likely to occur in incremental steps, possibly starting with an antisatellite or air defense weapon and, subsequently, expanding to deployment of a BMD system. If Moscow adheres to its long-standing pattern of evolutionary improvement to existing capabilities, it is likely that deployments of directed-energy weapons will occur in addition to those of more conventional defensive technologies. Indeed, the Soviets will probably co-locate SAM and ABM defenses with directed-energy antiaircraft or BMD weapons, thereby providing redundancy, robustness, and greater reliability against offensive countermeasures.

Soviet concerns in developing their own defenses focus on two types of strategic scenarios: a full-scale strike by U.S. forces against both Soviet forces-in-being and the elements of military potential; and a limited attack against strategic forces and/or C[3] centers. In discussing the missions of the air defense forces, Marshal G. V. Zimin identified five major mission areas for U.S. and NATO nuclear forces.

- To wreck or weaken significantly the second-strike forces and simultaneously to ensure the security of their government by destruction of ballistic missiles on the launch pads (in their silos), strategic bombardment of airfields, missile-carrying submarines in their base areas.
- To undermine the political and economic base of the state and crush the will of the people and their armed forces to resist by destruction of major political and industrial centers and areas and other important military and state objectives.
- To disrupt the systems of administrative and military control by the state and armed forces by strikes on command and control centers, communications means, and communication networks.
- To guarantee the means of attack by a breakthrough in the system of defense and access to their principal targets by suppression of the forces and means of the PVO.
- To complicate the organized employment of the armed forces in the conflict by direct action against principal groupings of ground forces and naval forces.[77]

Although this would certainly be a massive attack, it would not be simple "city busting." The Soviet description suggests that U.S. weapons would be directed at relatively discrete military, administrative, or control targets, many of which would be located in or near population centers. Zimin's analysis suggests the possibility that the Soviet Union might seek to defend elements of the U.S. target inventory, even against a major attack. Potential targets to be defended could include strategic nuclear forces, military-political C[3], key industrial sites, PVO forces, and even elements of the ground and naval forces. Defending the PVO forces would have the advantage of enabling those forces to attrite additional incoming offensive weapons.

A limited nuclear attack, as the Soviets envision it, would involve strikes against strategic nuclear forces, C[3] nodes, possible air defenses, and theater nuclear and conventional forces. The Soviet interpretation of the U.S. limited nuclear war strategy emphasizes not only the preemptive counterforce nature of the strategy but the political goal behind it—to ensure that the U.S. homeland remains secure by destroying or paralyzing Soviet retaliatory capabilities.[78] Recent Soviet statements have claimed that the U.S. goal is to conduct strikes sufficiently large to damage Soviet strategic and theater military capabilities while withholding strikes on large urban-industrial complexes, thereby placing before the Soviet leadership the difficult decision of acquiescence or escalation to all-out nuclear war.[79] U.S. strikes, it is implied, need not disarm the USSR, if withheld U.S. nuclear capabilities were sufficient to still threaten an assured destruction response and if damage to the USSR were limited.

What might Moscow choose to defend? Defense of high-value military targets against limited attacks may be an increasingly attractive option to Soviet strategic planners. Traditionally, the Soviet Union has not shown much interest in the defense of its ICBMs and other strategic assets. Instead, it would appear to rely on launch on warning and on the limited counterforce capability of U.S. strategic forces. However, as U.S. forces become capable of threatening a larger portion of the Soviet strategic force posture, and as U.S. forces become less vulnerable to Soviet preemption, strategic defense may come to be viewed by Moscow as a more viable means of preserving the effectiveness of the Soviet deterrent. The next generation of Soviet ICBMs will include one or more mobile systems. However, Moscow will continue to deploy a large number of silo-based systems well into the twenty-first century. Protecting this investment from attack would be desirable, if not essential.

Alternatively, the Soviets could attempt to expand on the current Moscow ABM and defend cities and value targets. Even if a Soviet preemptive strike were only partially successful, the United States would be left with a critical shortage of warheads with which to accomplish the Single Integrated Operational Plan (SIOP). Defenses of cities would challenge the United States to expand scarce resources in attempting to penetrate, thereby providing virtual protection to either Soviet military power or its urban-industrial missile and postattack reconstruction capabilities from destruction. Given the protected structure of Soviet offensive forces, this type of defense could severely hamper U.S. ability to effectively deter Moscow, and, in the event of war, ensure the United States' defeat.

Historically, the Soviet Union usually has not competed directly with the United States in the design of its own force posture. However, the Soviets have sought to match key U.S. technological achievements. Instead, Moscow has tended first to pursue options eschewed by or unavailable to the United States and later to close existing gaps in its force posture once the technology becomes available. Direct competition with the United States in the development of space-based defenses would constitute an extremely high-risk venture by the Soviet Union. It is more likely that Moscow would respond to the U.S. SDI with a combination of changes to its offensive force posture, development of additional limited strategic defenses, and research into counters to the types of defenses selected by the United States. Moscow certainly would like to deploy a preclusive defense, if this were possible. However, reliance on a single technology or type of system is viewed as inherently dangerous. Hence, the Soviets are likely to pursue strategic defenses incrementally, adding additional layers and new technologies to supplement existing, reliable capabilities.

Changing Soviet Doctrine and the Future

Soviet military doctrine and strategy are in the midst of a period of change. There is evidence that the Soviet military is in the process of writing a new set of field regulations, the first since the early 1960s. These regulations will define the manner in which the Soviet military will conduct wars, how it will train its personnel, and the type of forces it will create and sustain. The timing of this activity is important because it comes in the midst of a growing debate within the USSR on the nature of modern warfare, the role of nuclear weapons, the impact of new technologies on forces, strategy and tactics, and the size and composition of the Soviet Armed Forces.

The optimum Soviet strategy is to retain its present advantages in offensive forces while improving its ability to defend against damage to the homeland. Timing is crucial to attainment of this preferred posture. If the Soviets can pace U.S. defensive deployments with offensive counters while also expanding Soviet defenses, then the USSR may find itself in a better strategic position despite the deployment of defenses by the United States and its allies. The United States must either outpace Soviet offensive deployments or develop an offensive force posture capable of penetrating Soviet defenses.

It has been suggested that the Soviets might act aggressively against U.S. defensive deployments during the transition period. Although there are possible measures the Soviets could take, there are also plausible countermeasures. Moreover, the notion of a Soviet peacetime "attack" on U.S. military assets flies in the face of the lessons of history. Deployment of military capabilities as distinct from military actions with those forces has not been a cause of war, certainly not in the post–World War II era. Indeed, the continued presence of offensive nuclear forces in the transition would militate against the Soviet Union's willingness to risk an immediate catastrophe to prevent a future, theoretical, U.S. advantage.

The threat as viewed by Moscow is real, albeit not unexpected. The Soviets are well positioned to contest U.S. force posture changes. This is not to suggest that Moscow will be able to defeat the strategic defense program, but that it is likely to try. At the same time, Moscow also is aggressively pursuing its own strategic defense program. Soviet defenses alone may necessitate major changes in U.S. strategic forces if deterrence is to be maintained. This reaction, too, is not unexpected by Moscow. The competition, whether in defenses, offenses, or both as viewed by the Soviets will be enduring. The process is not uncertain, only the outcome. Even that, in the Soviet perspective, is only temporary. The SDI is merely one potential revolution on a path to a future replete with similar events.

Notes

1. See Roman Kolkowicz, "U.S. and Soviet Approaches to Military Strategy: Theory Versus Experience," *Orbis* 25, no. 2 (Summer 1981); Fritz Ermarth, "Contrasts in American and Soviet Strategic Policies," *International Security* 2, no. 2 (Fall 1978); Benjamin Lambeth, "The Political Potential of Soviet Equivalence," *International Security* 3, no. 2 (Fall 1979).

2. Leon Gouré, et al., *The Role of Nuclear Forces in Current Soviet Strategy*, (Coral Gables, Fla.: Center for Advanced International Studies, University of Miami, 1977), pp. 25–28.

3. N. V. Ogarkov, "Voyennaya Stratyegiya," *Sovetskaya Voyennaya Entsiklopediya*, vol. 7 (Moscow: Voyenizdat, 1978), p. 563.

4. N. N. Azovetsov, *V. I. Lenin i Sovetskaya Voyennaya Nauka* (Moscow: Izdatelstvo Nauka, 1981), p. 32.

5. See Mark E. Miller, *Soviet Strategic Power and Doctrine: The Quest for Superiority* (Bethesda, Md.: Advanced International Studies Institute, 1982).

6. N. V. Ogarkov, *Vsegda v Gotovnost Zashchite Otechestva* (Moscow: Voyenizdat, 1982), pp. 57–58.

7. See Michael J. Deane, *Strategic Defense in Soviet Strategy* (Bethesda, Md.: Advanced International Studies Institute, 1980); Sidney Graybeal and Daniel Gouré, "Soviet Ballistic Missile Defense Objectives: Past, Present and Future," in *U.S. Arms Control Objectives and the Implications of Ballistic Missile Defense* (Cambridge, Mass.: Center for Science and International Affairs, Harvard University, November 1–2, 1979).

8. A. T. Altunin, "Humanitarian Aims, Responsible Tasks," *Agitator Armiya i Flota*, no. 3 (February 1980):8.

9. G. Nekrasov, "Observer," Radio Moscow, February 6, 1977, cited in *FBIS (Soviet Union)*, February 17, 1977, p. AA-2.

10. Ogarkov, *Vsegda v Gotovnost*, p. 36.

11. G. V. Zimin, *Razvitiye Protivovozdushnoi oborony* (Moscow: Voyenizdat, 1976), pp. 4–5.

12. Ibid., p. 99.

13. D. A. Krutskikh, *Uchebno-Methodicheskoye Posobiye po Podgotovka Rukovoduyeshovo i Kommando-Nachal-stvuyushchevo Sostava Grazhdanskoi Oboroni* (Moscow: Voyenizdat, 1978), pp. 8–9.

14. N. V. Ogarkov, "The Defense of Socialism: Experience of History and the Present Day," *Krasnaya Zvezda,* May 9, 1984, p. 2.

15. S. Akhromeyev, "The Superiority of Soviet Military Science and Military Act—One of the Most Important Factors of Victory in the Great Patriotic War," *Kommunist,* no. 2 (February 1985):60.

16. V. Kulikov, "May 9—Victory Day: For the Sake of Peace on Earth," *Izvestiya,* May 9, 1984, p. 1.

17. Ogarkov, The Defense of Socialism, p. 3.

18. M. Kir'yan, *Voyenno-tekhnicheskiy Progress i Vooruzhenniye Sili SSSR* (Moscow: Voyenizdat, 1982), pp. 256–258.

19. Ogarkov, *Vsegda v Gotovnost,* p. 44.

20. A. A. Grechko, *Vooruzheniye Sili Sovetskovo Gosudarstva,* 2nd ed. (Moscow: Voyenizdat, 1975), p. 193. See also D. A. Volkogonov, *Voina i Armiya* (Moscow: Voyenizdat, 1977), pp. 195–213; Kir'yan, *Voyenno-tekhnicheskiy Progress.*

21. D. F. Ustinov, "A Victory Which Dispels the Myth and Illusion," *Pravda,* May 9, 1984, p. 2.

22. Akhromeyev, "The Superiority of Soviet Military Science," p. 61.

23. A. Bovin, "Fantasies and Realities," *Izvestiya,* April 21, 1983, p. 3; P. Zhilin, "Whence the Threat to Peace," *International Affairs* (Moscow), no. 5 (May 1982):8–14; V. Falin, "Facts and Fantasy," *Izvestiya,* April 10, 1985, p. 5.

24. S. Sokolov, "Answers to Questions by a TASS Correspondant," TASS, May 4, 1985, cited in *FBIS (Soviet Union),* May 6, 1985, p. AA-1.

25. G. A. Trofimenko, "Washington's Strategic See-Saw," *SShA: Ekonomika, Politikia, Ideologiya,* no. 12 (December 1980):55. Also G. A. Trofimenko, *Voyennaya Doktrina SShA* (Moscow: Znaniya, 1982), pp. 39–42.

26. V. M. Bondarenko, *Sovremmenaya Nauka i Voyennovo Delo* (Moscow: Voyenizdat, 1976), p. 132.

27. V. Aduyevskyu, "Space Must Remain Peaceful," *Kommunist,* no. 15 (October 1983):117; Yu. A. Zamoshkin, "The Nuclear Threat and The Factor of Fear," *SShA: Ekonomika, Politika, Ideologiya,* no. 3 (March 1985):3–6; A. Bovin, "Yet Another Washington Lie," *Izvestiya,* June 18, 1985, p. 5.

28. See Daniel Papp, "Soviet Perceptions of the Strategic Balance," *Air University Review* 32, no. 2 (January-February 1981):77–89.

29. A. Arbatov, *Voyenno-Strategicheskii Paritet i Politika SShA* (Moscow: Politizdat, 1984), pp. 28–33.

30. S. Sokolov, "Answers to Questions of TASS Correspondent," *Krasnaya Zvezda,* May 5, 1985, p. 3.

31. L. Semeyko, "In Pursuit of Superiority: The Dangerous Nature of the Development of U.S. Military Doctrine," *Krasnaya Zvezda,* August 5, 1983, p. 2.

32. For example, see "On the United States So-Called 'Strategic-Defense Initiative,'" *Izvestiya,* January 25, 1985, p. 5.

33. Moscow Round Table, April 19, 1985, in *FBIS (Soviet Union),* April 22, 1985, pp. 11–12.

34. A. Krasov, "For a Peaceful Cosmos," *Izvestiya,* June 26, 1982, p. 3.
35. Ogarkov, *Vsegda v Gotovnost,* p. 36.
36. P. Agafonov, "The Pentagon's Space Ambitions," *Krasnaya Zvezda,* March 31, 1981, p. 3.
37. Tolubko, *Kommunist,* no. 3, February 1985, pp. 49–63.
38. Akhromeyev, "The Superiority of Soviet Military Science"; Sokolov, "Answers to Questions," May 6, 1985.
39. Ferene Szues, "Capitalism in the 1980s: Crisis and Search for Solutions"; *Nushadscrel,* February 23, 1985, p. 9, in *JPRS World Report (Arms Control),* April 29, 1985, p. 9.
40. V. Chernyshev, "Sinister Deception," *Krasnaya Zvezda,* April 4, 1985, p. 3. Also see Yu. Zhukov, "Treacherous Deception," *Pravda,* March 20, 1985, p. 4.
41. Yuri Kolusov, *Struggle for Peaceful Space* (Moscow: Mezhdunarodniye Otnosheniye, 1984), p. 3. Also V. Vinogradov, "The Pentagon's Yearning for Space," *Krasnaya Zvezda,* April 6, 1984, p. 3; Falin, "Facts and Fantasy."
42. V. Kazakov, "Washington's Star Wars Ambitions," *Novoye Vremya,* no. 14 (April 1984):21.
43. V. Bogachev, "U.S. Militarists Seek to Move the Arms Race Into Outer Space," TASS, July 6, 1984, cited in *FBIS (Soviet Union),* July 7, 1984, p. AA-1.
44. For example, see G. S. Khozin, "America's Space Program at the Start of the 1980s," *SShA: Ekonomika, Politika, Ideologiya,* no. 8 (September 1982):33–34; A. Arbatov, *Voyenno-Strategicheskiy Paritet i Politika SShA* (Moscow: Politizdat, 1984); Agafonov, "The Pentagon's Space Ambitions."
45. Bovin, "Fantasies and Realities," p. 2; N. Chervov, "Defense or Attack—On Reagan's Anti-Missile Decision," *Pravda* (Bratislava), April 29, 1983, p. 2.
46. Arbatov, *Voyenno-Strategicheskii Paritet.*
47. S. Menshikov, "The Space Games of the Military Complex," *Pravda,* February 18, 1985, p. 6.
48. Committee of Soviet Scientists for Peace Against Nuclear Threat, *Strategic and International-Political Consequences of Creating a Space-Based Anti-Missile System Using Directed Energy Weapons* (Moscow, 1984), pp. 22–25.
49. A. Kokoshin, "Space-Based Anti-Missile Defense: Illusions and Dangers," *Moscow News,* no. 25 (June 1984):8.
50. Committee of Soviet Scientists, *Strategic and International Political Consequences.*
51. Y. Velikhov, "Protecting the Future Star Wars Illusions," *Sovetskaya Russiya,* April 21, 1985.
52. "News and Views," *APN Daily Review,* June 4, 1985, pp. 1–2 in *FBIS (Soviet Union),* June 23, 1985, p. 54.
53. G. Arbatov, "Interview with Correspondent Fujita," *Yomiuri Shimbon,* in *FBIS World Wide Report (Arms Control),* April 16, 1985, p. 20.
54. Solokov, "Answers to Questions," May 6, 1985.
55. A. Krasikov, "For a Peaceful Cosmos," *Izvestiya,* June 16, 1982, p. 2.
56. S. Akhromeyev, "The ABM Treaty Is an Obstacle to the Path of a Strategic Arms Race," *Pravda,* June 4, 1985, p. 4.

57. Edward Teller, "Science and Technology in the Strategic Defense Initiative," *DS2003+* (April-May 1985):21.

58. James A. Oberg, "Andropov's Orbiting Bomb," *Reason* (December 1983):25–30; U.S. Department of Defense, *Soviet Military Power, 1985* (Washington, D.C.: U.S. Department of Defense, 1985).

59. N. Chervov, "USSR Has Antisatellite System," *Frankfurter Rundschau,* cited in *FBIS (Soviet Union),* May 30, 1985, p. AA-6.

60. I. Lukin, "Conference of Experts on the USA," *SShA: Ekonomika, Politikia, Ideologiya,* no. 10 (October 1984).

61. See Daniel Gouré and Gordon H. McCormick, "Soviet Strategic Defense: The Neglected Dimension of the U.S.-Soviet Balance," *Orbis* 24, no. 1 (Spring 1980).

62. William Gwertman, "New Soviet Combat Aircraft: Quality With Quantity," *International Defense Review,* no. 1 (January 1984):35–38; U.S. Department of Defense, *Soviet Military Power,* pp. 36–39.

63. U.S. Department of Defense, ibid., p. 39.

64. Ibid., pp. 32–33.

65. Philip J. Klass, "U.S. Scrutinizes New Soviet Radar," *Aviation Week and Space Technology* (August 22, 1983):19–20; Clarence A. Robinson, Jr., "Soviets Accelerate Missile Defense," *Aviation Week and Space Technology* (January 16, 1984):14–15.

66. *Soviet Non-Compliance With Arms Control Agreements* (Washington, D.C.: The White House, February 1, 1985), p. 8; R. Jeffrey Smith, "Experts Condemn Soviet Radar," *Science* (March 21, 1985):1442–1444.

67. Soviet Radar Tied to Satellite-Killer System," *Chicago Tribune,* March 12, 1984, p. 2.

68. U.S. Department of Defense, *Soviet Military Power,* p. 33.

69. Robinson, "Soviets Accelerate Missile Defense," p. 15.

70. "USSR Developing Complete New Range of Ballistic Missiles," *International Defense Review* no. 9 (September 1983):1193; "Soviets Test Missile Reloads," *Aviation Week and Space Technology* (August 29, 1983):18.

71. Robinson, "Soviets Accelerate Missile Defense," p. 15; Clarence A. Robinson, Jr., "Soviet Making Gains in Air Defense," *Aviation Week and Space Technology* (April 1, 1984):22.

72. Stephen M. Meyer, "Soviet Military Space Programs and the 'New High Ground,'" *Survival* 25, no. 5 (September/October 1983):214–215; Charles S. Sheldon, III, *United States and Soviet Progress in Space: Summary Data Through 1980 and a Forward Look* (Washington, D.C.: Congressional Research Service, January 15, 1981).

73. Nicholas A. Johnson, "Soviet Satellite Reconnaissance Activities and Trends," *Soviet Aerospace Almanac, 1981* (Air Force Magazine, April 1981), pp. 90–92.

74. Thomas O'Toole, "Soviet Plan to Construct Space Base," *Washington Post,* December 27, 1982, p. A1; Rolf Engel, "Soviet 'Super Booster' ready for Relaunch?" *International Defense Review,* no. 3 (March 1982):279–297; "Soviet Shuttle a Carbon Copy?", *Aerospace America* (July 1984):1.

75. U.S. Department of Defense, *Soviet Military Power,* pp. 34–35, 38–39.
76. I. I. Anureyev, *Oruzhiye Protivo-Raketnoi i Protivo-kosmicheskoi Obornoy,* (Moscow: Voyenizdat, 1971), pp. 199–221.
77. Zimin, *Razvitye Protivovozdushnoi Oborony,* p. 87.
78. A. Slobodenko, "The Strategy of Nuclear Adventurism," *Mezhdunarodnaya Zhinz,* no. 1 (January 1981):30–31; L. Semeyko, "Gambling on a First-Strike Potential," *Krasnaya Zvezda,* August 8, 1980, p. 3; P. G. Bogdanov et al., *SShA: Voyenno-Strategicheskiye Kontsepsii* (Moscow: Nuaka, 1980), pp. 61–65.
79. Ye. Rybkin and S. Kortunov, "Military Parity as a Factor in Security," *Mezhdunarodnaya Zhizn,* no. 8 (August 1982):48.

9

Soviet Strategic Defense and Military Space Systems

Richard E. Thomas

The Soviet Union appears to be relatively far along in specifying the architecture of its strategic defense system, in developing the technologies that will undergird that system, and in deploying some of its elements. The Soviets understand the advantages of layered defenses and the problems of effective battle management. However, all this activity is but a part of a larger effort that envisions a full range of military operations in space, an effort that will seek to provide the Soviets supremacy in that vital theatre and one that has been underway for at least two decades.

In the autumn of 1984 I visited several organizations in Western Europe where I spoke with academicians engaged in strategic and international policy research. These discussions occurred roughly eighteen months after President Ronald Reagan's speech describing the Strategic Defense Initiative (SDI)—by the time the reactions to the new U.S. program were in full flower in the United States, Western Europe, and the Soviet Union.

The Western European response I encountered can be verbalized in the following fashion: "If you Americans persist in developing the SDI, you will *force* the Soviets to do the same thing." My response was, "Have you not read Sokolovskii and the writings of other Soviet military leaders that describe Soviet strategic defense and military space programs?" I was uncertain whether my NATO colleagues had read the literature or had ignored its message.

It is clear from the papers that undergird and describe Soviet military doctrine and from the known weapons development programs active in the USSR that the Soviets have been considering, planning, and implementing strategic defense and military space systems for some time.

These activities have involved efforts to forecast the technologies and organizational structures required for such systems. These efforts, in turn, have led to massive research programs in the Soviet scientific and technical community to evolve the needed technologies as well as programs to develop and deploy the weapons systems.

One increasingly hears the term *Soviet SDI* used to describe some of these activities, but that choice of words is unfortunate. It implies that the Soviet SDI will be a mirror-image of or symmetrical to the U.S. SDI. Although the Soviets already have and will continue to develop elements of a system to counter ballistic missiles, it will have a different configuration, be prioritized differently in its evolution, and include different weapons than will the U.S. SDI.

My purpose here is to analyze some of the Soviet literature that reflects Soviet views on ballistic missile defense (BMD) systems and on the large arena of military space operations. The focus of this endeavor will be the Soviet doctrinal literature that tends to describe the features of policy decisions as they relate to both organizational and technological requirements. These decisions arise out of the consideration of several alternative approaches that in the 1960s appeared on the pages of various Soviet military journals, the "ordered ferment" as described by Dziak.[1] Although these days one sees in the literature less-direct evidence of the trends of Soviet military doctrine, the results of those decisions are clear in Soviet military textbooks and in scientific documents, which are evidence of the central direction that is the hallmark of the Soviet system.

The characteristics of the weapons systems that currently are part of the Soviet strategic defense systems will not be reviewed extensively here—the reader can obtain that information from other readily available sources. Instead, I want to emphasize the general parameters of Soviet thinking about military space and strategic defense systems so that a conceptual basis can be established for considering future systems. If we are to be successful in such an endeavor, we must think as the Soviets do.

Soviet Versus U.S. Strategic Defense

The differences in the two strategic defensive systems, U.S. and USSR, derives from fundamentally different philosophies. "In the Soviet system military doctrine drives technology, while in the U.S., it's the other way around." Within the limits typical of all sweeping generalizations, the statement is true, but the Soviets do consider technological alternatives before making doctrinal decisions just as leaders in the United States consider geopolitical and defense issues before making technological decisions. Additionally, the United States typically has used technology to balance the numerical edge in personnel and weapons held by the Soviets. In recent years the Soviets have become more concerned about the impact on the military balance of the technological initiative, the rate of development of weapons technologies, and the quantity/quality debate.

As a result of these differences, the Soviets emphasize ground-based systems for missile defense—command and control concerns are a major impetus in the design of Soviet weapons. Technologies are used differently, such as the notion of antiballistic missile (ABM) systems deployed on submarines. Although Soviet weapons may appear to be similar to their U.S. counterparts, we must resist the temptation to describe those systems in a U.S. framework. We must be prepared to see a uniquely Soviet system.

In 1952 when I was employed at U.S. Air Force Air Technical Intelligence Center (now the Foreign Technology Division), I analyzed Soviet advances in various facets of missile technology, particularly rocket engines, and predicted the appearance of Soviet intercontinental ballistic missiles (ICBMs) in 1957.[2] The death of Stalin in 1953 seems to have paved the way for the Soviet decision to leapfrog the air-breathing weaponry that was a large part of U.S. thinking at that time (the early cruise missiles Matador, Snark, and Navaho, which were to augment a predominantly manned bomber force). Soviet strategists, freed of the repression of the Stalinist era, began consideration of the fundamental tenets of the revolution in military affairs, which had as its main theme the impact of the nuclear-tipped ICBM on the world military and political framework.

The Soviet penchant for dialectical thinking led military thinkers in a natural way to begin early consideration of ways to defend against ballistic missiles. Indeed, some of that thinking evolved out of British ideas for defense against the V-2 during World War II. While some in the West saw the ICBM as the ultimate weapon, the Soviets steadfastly believed that no such weapon could ever exist, that any weapon could be countered in some way; nor did the Soviets limit themselves to the consideration of ABM systems. Having perceived the U.S. space program as a military threat, Soviet military thinkers in the early 1960s were considering ways to defend against military vehicles deployed and operated in space. Many of these writings became a part of the first edition of Marshal V. D. Sokolovskii's extensive work, *Soviet Military Strategy,* first published in 1962.[3]

In that work Sokolovskii stated that the contemporary theater of military operations extended far beyond "the firing range of weapons" and included entire continents, oceans, and substantial regions of the atmosphere and space. His was the first influential voice in the USSR to describe the need for the development of military space systems. The group of bright military minds that were part of the Sokolovskii circle in the 1960s was innovative in its conceptualization of the missions, weapons, and operations associated with what we today would call strategic defense. These strategists at that time, however, did not see space-based systems as integral to defense against ballistic missiles. Instead, "space-defense means" were needed to defend against hostile actions by "space apparatuses."[4] Aircraft and missile defense were considered the responsibility of ground-based units.

True to their commitment to the doctrine of combined arms, the Soviets were convinced that such operations would have to be integrated "in accordance with a single plan" and would involve "a single process of combatting the air-space forces of the enemy."[5] A major focus of their concern was the issue of command and control, the need for accurate information on targets, and the need for very rapid decisions. "On the whole the aerial-space defensive operations are characterized by great range, highly dynamic battles in encounters with the enemy, sharp changes in the situation and combat in conditions of extremely limited time when the outcome of combat actions is decided by minutes and even seconds."[6] Time constraints would demand the use of automated systems to control or to assist in controlling operations.

Interestingly, the limitations on such activities led to the conclusion that some decentralization was preferred so that tactical units could undertake actions in response to local conditions or in the event of a loss of communications with a central command. This overriding concern with command and control (C^2) issues suggests that those issues would constitute a major parameter in the establishment of the architecture of Soviet exo/endoatmospheric defense systems.

Soviet military doctrine deals with organizational as well as other matters, and in 1962 Sokolovskii described the need for two new structures to supplement the PVO command whose task was conventional aircraft defense. The mission of a new entity, PRO, was to be ballistic missile defense while another new organization, PKO, was to handle space defense. The integration of PVO and PRO was a natural result of the attack sequence expected by the Soviets: ballistic missiles, cruise missiles, tactical and carrier-launched aircraft, and strategic bombers.

The PRO mission was to be early detection and identification of missiles and warheads in an environment of decoys and electronic countermeasures (ECM), followed by interception and destruction of the warheads at a safe altitude and distance. Sokolovskii mentions "air defense (PVO) ships." The exact meaning of this statement is not clear, but other Soviet writers have discussed the possibility of installing ABM systems on both surface ships and submarines.[7] Soviet submarines are credited with an ability to mount offensive actions from polar regions. Although the conduct of both ballistic and cruise missile defense operations would require some unique weapons characteristics, this defense must be regarded as a definite possibility for the Soviet military.

The second edition of Sokolovskii's *Military Strategy,* published in 1963, contained the statement, "It is assumed that space weapons would occupy an important place in the future."[8] However, the third edition, published in 1968, was substantially modified and enlarged in dealing with military space operations. Keep in mind that this was the time of the U.S. Apollo program and President John Kennedy's public challenge to the Soviets to a race to the moon. The Soviets saw substantial military overtones to the U.S. effort and predicted U.S. military bases on the moon and in space.

Harriet Fast Scott, in her analysis of segments of Sokolovskii's book, offers a worthwhile comment on the third edition.

> The changed treatment the Soviets have given to the possibility of military use of space between this edition and the two previous editions is most interesting. The earlier editions had considered military use of space in Chapter VI entitled "Methods of Conducting Warfare" in a section on "Problems of Using Space for Military Purposes." Although the examples given were from United States sources, the editors did acknowledge that "Soviet military strategy takes into account questions as the use of outer space and aerospace vehicles to strengthen the defense of the socialist countries." The space discussion concluded with the statement, "The facts stated are evidence that the American imperialists have taken the path of direct use of space to realize their aggressive intentions."
>
> Later in Chapter VI, the Soviet editors are very careful to give an impression of Soviet supremacy in space. Here, the following addition is found, ". . . it must be assumed that in the near future, radical corrections will be able to be introduced into this system as a result of the incorporation of various cosmic means. All this in turn conditions the nature of a future war, the methods of waging it and the principles of organization of the armed forces." This could be interpreted that the Soviets are taking a very active interest in the military use of space while practicing traditional Russian secrecy. In another publication Soviet military spokesmen have noted that the creation of a weapon that is new in principle and secretly nurtured in scientific research bureaus in constructors collectives can in a short time sharply change the relationship of forces. The *surprise* appearance of one or another type of weapon is advancing as an essential factor, especially in contemporary circumstances.[9]

Starting from these early considerations, Soviet strategic thinkers established in the late 1960s and early 1970s an ever more complex framework of missions, technologies, and weapons needed for the tasks associated with ballistic missile defense and military space operations. As is usually the case with Soviet military thinkers, they appear to have attempted forecasting the rate of technologies and weapons development and to have keyed their plans to those forecasts.

Given the substantial Soviet reaction to the SDI and the considerable Soviet efforts through political and diplomatic means to stop or slow that program, it seems that the Soviets perceive the SDI has altered the forecasted rate of technological development and that as a result some weapons might become operational considerably sooner than the Soviets had anticipated. Therefore, they may be forced to alter and reprioritize their weapons development programs.

In 1968 the military space missions described by the Sokolovskii group were reconnaissance or service missions, such as launch detection of both land-based and submarine-based ballistic missiles, the monitoring of nuclear weapons tests in the atmosphere, and damage assessment of targets attacked with nuclear weapons and military satellites for communications, navigation, and weather reconnaissance. The currently operational Soviet antisatellite (ASAT) weapon was probably on the drawing boards at that time (initial testing began in the early 1970s). That weapon was apparently directed at U.S. reconnaissance satellites—Sokolowskii stated that because those vehicles are constrained to move in definite orbits, they are vulnerable and can be destroyed by "anti-space defense weapons."

At the same time other Soviet authors[10] were considering the problems posed by MIRVed missiles, such as the need to destroy the "carrier" (booster) vehicles before the warheads were deployed. In ICBM parlance, the notion of boost-phase intercept clearly had been introduced, although Soviet writers may have been focusing on weapons that would deploy warheads from space, vehicles such as the Soviet fractional orbital bombardment system (FOBS). The need to be able to intercept individual warheads also was noted. Thus, the concept of layered defensive systems appeared early in Soviet thinking.

Battle management and other command, control, communications, and intelligence (C^3I) issues also appeared in the early Soviet literature,[11] with the call for systems that could not only detect but also identify and track hostile objects moving through space. This was to be accomplished by both earth-based and space-based sensor systems feeding information to a main control center on earth. However, a backup control center was to be located in space!

Although in 1966 the Soviets established the requirements for long wavelength infrared (LWIR) systems to operate in the wavelength range from 0.76 microns (the end of the visible spectrum) to 500 microns (!), the tasks of detection and tracking were initially to be accomplished using long-range radar units. Trajectory data were to be used in order to predict the targets under attack and the time of the attack. The missile defense control center then would select the appropriate weapons for repulsing the strike. ABM systems would utilize missiles designed for both high-altitude and low-altitude interception of the warheads. Although the flight paths of land-based missiles could be accurately predicted, submarine-launched ballistic missiles (SLBM) posed special problems. However, the Soviets believed that even here reasonable predictions would follow from a careful analysis.

In all of their considerations of sensor systems, the Soviets worry extensively about *maskirovka,* the Russian word that generally means "camouflage" but that also connotes concealment and deception. Perhaps because of their own commitment to the study, development, and utilization of the techniques of *maskirovka,*[12] they believe the United States is similarly committed. The Soviets therefore must be prepared to handle such a threat.

In the Soviet view one of the best ways to accomplish the detection, identification, and tracking of ballistic missiles, space vehicles, and warheads is through the use of electro-optical devices.[13] The units have two major advantages. First, they can be used in passive as well as active modes, and, second, they are generally smaller and more compact than an equivalent microwave device. In theoretical studies of multipixel LWIR detector systems, the Soviets have put special emphasis on the determination of the sensitivity needed to detect a target against a background that might be purposely cluttered. For the detection of radiation from rocket engine exhaust gases, space-based sensors must operate in the infrared wavelengths that are transmitted through the atmosphere—namely, 3–5 microns, 7–9.5 microns. The spectral characteristics of rocket exhausts constitute a "signature" that can be useful in identifying the source of the missile, but the Soviets noted that chemical additives to the propellants can alter the signature. For the task of tracking and targeting, laser radars are preferred over both passive electro-optical systems and conventional microwave radars because of the increased activity needed for "determination of the coordinates of fast moving targets and, secondarily, for guiding anti-missiles to a target along a laser beam."

As part of their intercosmos program the Soviets have installed laser radar stations in friendly countries around the world, including one in Santiago de Cuba, an excellent vantage point for gathering data on U.S. vehicles launched down the Atlantic test range. Soviet literature places great emphasis on gathering and computerizing the reflected and emitted radiation of boosters, missiles, and spacecraft. These data are to be gathered at several wavelengths, presumably to forestall enemy *maskirovka* efforts.

After the ABM treaty was signed in 1972, Soviet discussions of such systems were somewhat muted, but Soviet literature in the years between Sokolovskii's landmark studies and the ABM agreement provides an interesting perspective on Soviet views on space and missile defense systems and on military space operations in general. As is traditionally the situation, much of these discussions are based on U.S. systems. However, one can discern certain views that most likely reflect Soviet thinking about Soviet systems.

One of the more important documents of that period was authored by Major Gen I. I. Anureev,[14] who was a professor at the Academy of the General Staff. The writings of the members of the faculties of the higher military academies of the Soviet Union constitute an important source of information for Western analysts. The top military academies have "think tanks" associated with them wherein analyses of advanced military strategies, tactics, and weapons systems are conducted. These studies form the basis of the doctrinal statements issued by military leadership and later of the textbooks used in the academies.

It is worthwhile to note some of the more significant elements of Anureev's book. These not only describe those facets of ballistic missile and military space operations the Soviets consider most important, but also reveal some of the unique, and to many in the West, strange ways in which the Soviets evaluate U.S. plans and activities. Let me cite an example. Much of the U.S. population, including those of us who were involved in the Apollo program of the 1960s, know that program was designed for only one mission—to fulfill President Kennedy's charge to place a man on the moon and return him safely to earth. Anureev saw it differently.

> The developed ship, almost without modification, can be used for inspecting unidentified satellites or for servicing spacecraft. Moreover, the task has been set of using these ships for reconnoitering ground objects from space, as well as manned satellite interceptors.
>
> For orbital interception they intend to use the lunar module which in an orbital flight is capable of rendez-vousing with virtually any spacecraft while having astronauts leave the spacecraft opens up broad opportunities for examining these objects.[15]

Anureev's study of antimissile and space defense weapons reflects the then generally held view that offensive weapons technologies were superior to defensive ones and that advances in sensor systems and computers would be required in order for the prospects for effective

ABM systems to improve. He recognized the central issue of discrimination—detecting a warhead in a "cloud of decoys"—but he believed that systems of sensors operating on multiple wavelengths and utilizing both microwave as well as active and passive optical systems would be equal to the task. Although laser weapons were mentioned, the clear emphasis was on kinetic-energy kill mechanisms, and there was little doubt that such vehicles could be designed and developed. Land-based systems were preferred by Anureev for all phases of ABM activities. Once the warheads had been identified, then the primary problem was the time constraints on the command and control elements of the system. Computer-assisted decisionmaking techniques clearly were required.

Anureev described several penetration aids that would have to be considered in designing the ABM system. Dipoles or chaff might be dispensed to fly freely in space or tethered to a warhead or a decoy. The electromagnetic pulse from a high-altitude nuclear burst temporarily might disable electronic devices in the defensive system. The detectability of a warhead could be reduced by using radar-absorbing material in its structure and by pumping cold gas into the plasma sheath that forms around a warhead during reentry. Metallized balloons could be configured to look like warheads. Electronic jammers could disrupt radar devices. Warheads could be designed to maneuver in the atmosphere or to reenter at high angles in order to reduce the intercept time. Because these techniques were expected to be used by the offense, it is likely that Soviet missile defense systems would be designed to counter them. Again, reliance was to be placed on multispectral detection, identification, tracking, and kill assessment devices.

As is well known, the ABM treaty permitted the United States and the USSR each to have one operational system. The United States initially deployed, then dismantled, its system around the ICBM silos in the northern part of the nation. If that system had been considered effective, it would have guarded against a Soviet first strike. However, it was not able to handle multiple warheads accompanied by a multitude of decoys.

The Soviets on the other hand have deployed and steadily modernized the ABM system that encircles Moscow.[16] Although they accuse the United States of preparing a first strike, they choose not to defend their ICBM launch sites but prefer to assure the survival of the central elements of the Communist party housed in Moscow. One wonders who is preparing the first strike.

Several other aspects of Anureev's discussions of military space operations deserve mention. Other analysts[17] have noted his concern about access to space and its impact on the military potential of the region. The importance of heavy lift boosters and reusable spacecraft are described and clearly have influenced the Soviet space program up to the present time.

Anureev was one of the first authors to describe the potential for the full range of military operations in space—operations encompassing offensive as well as defensive, reconnaissance, and support activities such as those provided by navigation and communication satellites. In this connection he appeared to conceive of space as a theater of military operations (TVD) in the parlance of the Soviet military. This is an important, far-reaching concept that has since appeared in more recent Soviet writings. I shall return to it later.

The most important function assigned to space-based assets was reconnaissance, which included the identification and location of targets and the effectiveness of attacks on them, the monitoring of enemy activities on land and the oceans, particularly the detection of ICBM and SLBM launches. These systems were linked with communications satellites to perform a singularly vital function: the support of military operations on the earth's surface. With this type of mission prioritization, the early development of the Soviet ASAT system[18] was a very logical move. However, it seems clear that the Soviets do not rely exclusively on their ASAT to counter enemy space-based assets. Anureev stated that "space defense involves the creation of conditions on the earth's surface to impede the operation of military spacecraft."[19] This would seem to signify the use of *maskirovka* technologies on ground installations and to reduce the detectability of missiles during the launch phase, practices that are well developed and extensively pursued in the Soviet Union.

In addition, the Soviet literature discusses the use of ground-based weapons systems against satellites. These include ground-based lasers, high-power microwave devices, and direct-ascent missiles fired from either land bases, ships, or submarines. Many of these systems would rely on the comparatively low maneuverability of most current military satellites and would be most effective against those vehicles operating in low earth orbit.

Interestingly, the Soviets divide space into three spherical regions. Circumterrestial space is the zone ranging from 37 to 100 miles altitude where conventional aircraft cannot operate, but transatmospheric vehicles powered by scramjet engines and flying at hypersonic speeds could reach orbital velocities. Near space ranges from 100 to 310 miles, the outer limit of which is well inside the radiation belt. Outer space extends from 310 miles to about twice the earth-moon distance, approximately 558,000 miles. The Soviets clearly believe that a spacecraft's vulnerability reduces as its orbital altitude increases.

The capability to maneuver is one feature that the Soviets have found attractive in the shuttle.[20] In addition to its ability to move when being attacked—assuming, of course, that there is adequate time to sense the attack, determine its nature, and plan and execute the appropriate maneuver—an interceptor vehicle, especially with added supplies of propellant, can alter substantially its orbital plane as well as its other basic orbital parameters. Because military space vehicles are likely to be found in orbits ranging from equatorial to polar, including circular and quite elliptical orbits, and from low to geosynchronous altitudes, an interceptor based on the shuttle design would be militarily attractive. Although I am not familiar with any plan to use the shuttle for such a mission, the Soviets believe it could be used to capture, disable, or destroy military satellites.

The gliding range of a winged space vehicle means that it could reenter the atmosphere from a wide range of orbits and maneuver aerodynamically to its base. The Soviets have orbited and recovered a small winged space plane that is capable of carrying a single pilot and a variety of military equipment. In addition, the Soviets are developing their own shuttle, a craft having dimensions similar to the U.S. vehicle.

In considering the evolution of the Soviet strategic defense system, there are some other parameters that will influence its architecture, factors that have appeared in the Soviet military literature. The power requirements for space-based, directed-energy weapons are substantial and necessitate placing heavy power sources in orbit or the means of energizing some type of energy storage system from earth. Solar panels are large and vulnerable; therefore, military vehicles operating in space probably would use chemical or nuclear systems.[21]

It is likely that in the near term the power requirements for some strategic defense weapons would lead the Soviets to utilize ground-based systems wherein weight and size are relatively unimportant. Although the propagation of beam weapons through the atmosphere might produce some problems of intensity reduction because of absorption or the pointing of the beam, those issues are probably more tractable than is the power system matter. The Soviets have considered mounting lasers on mountain tops to reduce the thickness of the atmosphere through which the beam would propagate,[22] thereby reducing the amount of energy lost due to atmospheric effects. Although these systems might be too large and too heavy to place in orbit, they nonetheless could be portable and might be installed on ships and submarines.

This latter point becomes even more important when one considers the Soviet desire to defend the homeland as far from the borders of the USSR as is possible. Recent activities by Soviet submarines conducting operations in Swedish waters in conjunction with small tracked submersibles[23] could mean, according to some analysts, that the USSR intends, in wartime, to seize the so-called northern flank of Europe (Scandinavia) to assure control of the northern air and sea approaches to the USSR and as a base for an attack on the United Kingdom. The flight paths of some missiles and aircraft from land bases in the United States would taken them over Hudson Bay, Foxe Basin, and Baffin Bay areas. Therefore, it is reasonable to expect the Soviets to mount both aircraft and missile defense operations from ships and submarines in these areas.

Another factor that will affect the design of the Soviet strategic defense system concerns the allowable number of casualties in an attack on the USSR. U.S. estimates in the 1960s of the cost of terminal defense systems were based in some analyses on systems intended to protect 90 percent of the U.S. population. General Anureev noted that the excessive cost of such systems was one factor that drove the United States to sign the ABM treaty.

The configuration of the Soviet civil defense system suggests that the Soviets are prepared to accept a much higher percentage of casualties. That system appears to give highest priority to the protection of the leaders, the central elements of the Communist party, the armed forces, and the workers in essential industries. If this is correct, then the Soviet strategic defense system can be smaller than might otherwise be the case and could be focused in certain geographic areas.

The issue of command and control of military assets has been a matter of great concern in the Soviet Union for the past several years. The Soviet system has a heavy top-down character with little autonomy for local commanders; when used by Arab armies in the Middle East, the system has proven deficient. In spite of efforts to change the system to promote innovative thought, a greater degree of self-reliance, and limited independence from central authority, the Soviets have not yet achieved success. This was dramatically illustrated by the failure of their system in the destruction of KAL flight 007.

It may well be that nothing worries the Soviets so much as a failure in their command, control, communications, and intelligence (C^3I) system, the consequent lack of knowledge of events in battle, and an inability to control Soviet forces. Although space-based communication systems designed around laser devices will be used to augment conventional microwave units, the Soviets will put their greatest trust in earth-based systems at least for the near term.

Future Soviet Plans for Strategic Defense

It is clear from the positions stated in Soviet military doctrine and from the weapons systems development programs pursued to date that the Soviets have and will continue to develop systems to defend against ballistic missiles as well as other means for delivering strategic weapons on the USSR. They never have accepted the concept of mutual assured destruction but rather have acted on the belief that although ICBMs and SLBMs are formidable weapons, they, like any other weapon, can be countered. The Soviets have been planning, designing, developing, and deploying ABM systems for twenty years and, in parallel actions, are actively seeking to establish the capability to conduct extensive military operations in space.

The current Soviet strategic defense system probably contains three main elements:

1. A layered terminal defense system built around antimissile missiles activated by ground-based radar and space-based optical launch detection systems with battle management functions carried out using phased-array radar systems. The discrimination requirement will be achieved by using multispectral microwave and optical devices and atmospheric filtering of light weight decoys.
2. Boost-, postboost-, and midcourse phase intercepts will be attempted using ship- and/or submarine-based ABM or direct-ascent ballistic missiles.
3. The Soviet ASAT system and direct-ascent missiles, both land- and sea-based, will be used to attack U.S. space-based assets. Soviet land-based laser systems will be used to blind U.S. optical devices, and powerful microwave units can be used to disrupt electronic systems.

It is likely that as the Soviet strategic defense system evolves there will be continued reliance on ground-based systems, at least for the next several years. Based on the Soviet literature, however, there is no doubt that the Soviets will seek to establish a capability to conduct the full range of military operations in space.

Like the United States, the USSR presently has support satellites in orbit, and some analysts believe the Soviet Union has conducted military activities using the Salyut space station. In addition to the Soviet space shuttle, a very heavy lift booster is known to be in development.[24] When it becomes operational one of its first missions may be placing laser weapons in orbit. Although they may be low powered and comparatively ineffective as a weapon, the Soviet fear of being a second-rate power would be assuaged if the USSR were the first nation to place active weapons in space, a move the Soviets will say was forced upon them by the United States.

The development of the Soviet military space program probably will proceed in accordance with the evolving Soviet view of a space TVD, a space theater of military operations.[25] This concept is rooted in the writings of Sokolovskii and Anureev and has appeared frequently in more recent publications from the Soviet Union.[26]

The Soviet TVD is primarily a mechanism for effective command and control of a region of combat wherein several types of military operations are taking place. The contemporary description of a TVD on the surface of the earth is based on the delineation of specific geographic areas in which the full range of combined arms operations occur; however, the operations may be unique to that region. These operations include offensive, defensive, reconnaissance, and support activities, nuclear and conventional weapons, tactical and strategic targets.

The land TVDs ringing the Soviet Union include the Arctic Ocean TVD, far eastern TVD, the southern TVD encompassing the Middle East, a southwestern TVD focusing on southern Europe and northern Africa, the western TVD fronting on Western Europe, and a northwestern TVD that includes Scandinavia. Three other oceanic TVDs also have been defined. Some of these TVDs have no active forces assigned to them but rather serve as regions for Soviet military analysts to plan organizational and weapons requirements to support potential military operations in the area.

One presumes that a Soviet space TVD would involve the same types of concepts, although they would require different vehicles and weapons. Because the TVD is a framework for consideration of offensive as well as other weapons, Soviet thinking about military operations in space has transcended the views extant in the United States. Insofar as I am aware, U.S. thinking has encompassed only defensive, reconnaissance, and support operations. It indeed will be interesting to watch the Soviet space TVD concept unfold. The command structure for that organization probably already exists but is spread among several of the five Soviet military services. As military operations in space enlarge and diversify, we may see a new command structure emerge in the Soviet military to underscore the Soviet commitment to what may be described as a new "revolution in military affairs." Such a move would be similar to the emergence of the Strategic Rocket Forces from the "revolution" produced by ICBMs with nuclear warheads.

Some see the development of the space-based elements of strategic defense systems as heavily dependent on advanced technology, a view that results from descriptions of some elements of the Strategic Defense Initiative and from the Star Wars label pinned on it by the media. Although some of the more exotic weapons in the SDI will require research on and development of new scientific and engineering concepts, there is much that can be done with only modest advances in existing technologies.

This notion is very important when one attempts to forecast the path of weapons development in the USSR where evolution is more the rule than is revolution. Access to space may be enhanced by transatmospheric vehicles, powered by scramjet engines, operating from runways as do conventional aircraft but able to accelerate through the atmosphere to orbital velocities. The Soviets are studying these devices, but they for some time will continue to emphasize conventional rocket-propelled boosters using thoroughly tested and reliable engines. A weapon orbited by a rocket booster is just as effective as the same weapon placed in space by some more exotic means.

The Soviets have a variety of rocket-propelled, kinetic-energy weapons capable of intercepting targets in the atmosphere. The development of reaction controls to replace aerodynamic devices would allow these weapons to be used in space. They would constitute an effective intermediate space arsenal until the beam weapons technologies mature.

Again, this is not to say that early on the Soviets will not place a crude laser weapon in orbit. We must continue to bear in mind the prophetic words of Winston Churchill, "The Soviets want the fruits of war without war." The use of threat and intimidation is a very large part of the Soviet strategy. A space-based beam weapons system, however ineffective it might actually be, would provide the Soviets with substantial psychological leverage.

Military space systems will confront the Soviets with some technological requirements that will probably be most severe in the C^3I area. Soviet literature reveals great concern about the pace of events in modern warfare and the reduced time available for decisionmaking. As long as military operations in space involve rocket-propelled, kinetic-energy weapons, existing C^3I technologies are probably adequate. Distances in space are great, but even with hypervelocity vehicles, the times required to traverse those distances are of the order of minutes, quite adequate for sensor, communications, and decision systems to act.

That of course will change when effective beam weapons appear with their attack coming at the speed of light. Distance then would afford little time to act. Instead, C^3I devices must be able to sense the pointing and tracking operations underway by an enemy satellite; computers must predict the nature of the attack and assist in taking action to repel the attack or maneuver out of the way. The Soviets probably would choose to harden their systems to withstand the beam attack when it comes. Systems for pointing, tracking, and kill assessment probably will constitute the major technological hurdles confronting the Soviets on their way toward a space TVD encompassing large numbers of high-powered beam weapons.

The technological challenges that confront the Soviets in the matter of strategic defense systems and other advanced weaponry are well understood. Their literature is filled with papers assessing the relative merits of quantity and quality in weapons, and this debate has spawned a new sixth law of war, which was promulgated in late 1983. The development and shift in the methods of combat are dependent upon quantitative and qualitative changes in military technology and upon the level of moral and military qualities of military personnel. The Soviets have discussed the importance of the rate of development of weapons technologies and the importance in military affairs of "the technological initiative." Some of their military leaders appear to concede that through the SDI, the United States in fact has seized that initiative.[27]

Clearly, the Soviet leaders will try by political means to blunt the pace of development of U.S. military technology and will be making every effort to assess the impact of budgetary constraints, political changes, and public support in the United States. They recognize that in an all-out technological race with the United States, they are likely to lose. They also appear to realize, as do the Western Europeans, that the technological fallout from programs like the SDI will be enormous and that the gap that currently exists, particularly in the information processing technologies, most likely will widen. They cannot afford to let that happen.

Clearly, much of the military research and development currently underway in the United States focuses on the SDI and on determining the feasibility of developing and deploying effective systems to defend against both long- and short-range ballistic missiles. These efforts should be pursued if for no other reason than to assure that balance is maintained with the Soviets as they expand and refine their missile defenses.

The United States, however, must begin to consider, as the USSR has, the complete spectrum of military operations in space. This will require broadening the current planning and research programs beyond the SDI so as to provide the United States with weapons and support systems to meet the Soviet military challenge in space.

Notes

1. Dziak, John J., *Soviet Perceptions of Military Power: The Interaction of Theory and Practice* (New York: Crane, Russak and Company, 1981).
2. Thomas, Richard E. *Soviet Guided Missiles* (U.S. Air Force Technical Intelligence Center, 1952).
3. Sokolovskiy, V. D. *Soviet Military Strategy*, edited by Harriet F. Scott (New York: Crane, Russak and Company, 1984).

4. Kalugin, Y., "The Nature of Combat Operations of Air Defense Troops," *Voyennaya Mysl,* no. 1 (1968):43–52.

5. Ibid., p. 44.

6. Ibid., p. 44.

7. Stubbs, Kevin D., "Extending the Barrier: The Role of the Navy in Strategic Defense," in *Proceedings of a Conference on Soviet Military Personnel and Organizations,* Paper no. 8 (College Station: Center for Strategic Technology, Texas A&M University System, October 1985).

8. Sokolovskiy, *Military Strategy,* p. 60.

9. Ibid., pp. 47–48.

10. Kalugin, "The Nature of Combat Operations."

11. Ibid.; Provorov, K., "Missile and Space Offensive Weapons and the Problems of Countering Them," *Voyennaya Mysl,* no. 5 (1972):119–125.

12. Beaumont, Roger, *Maskirovka: Soviet Camouflage, Concealment and Deception* (College Station: Center for Strategic Technology, Texas A&M University System, November 1982); Smith, Charles L., and Joseph P. Zebrowski, *Evolution of Soviet Doctrine on Maskirovka of Fixed Installations* (College Station: Center for Strategic Technology, Texas A&M University System, October 1983).

13. Thomas, Richard E., *Long Wave Infrared Research in the Soviet Union* (College Station: Center for Strategic Technology, Texas A&M University System, March 1984).

14. Anureev, Ivan I., *Antimissile and Space Defense Weapons,* JPRS-57378, October 31, 1972.

15. Ibid., p. 94.

16. *Soviet Military Power* (Washington, D.C.: U.S. Government Printing Office, 1985).

17. Kipp, Jacob W. et al., *Soviet Views on Military Operations in Space* (College Station: Center for Strategic Technology, Texas A&M University System, October 1985).

18. *Soviet Military Power.*

19. Anureev, *Antimissile and Space Defense Weapons,* p. 203.

20. Volodin, A. "The Multiple-use Transport Spacecraft and Prospects for Its Development," *Voyennaya Mysl,* no. 2 (February 1973):50–55.

21. Sagdeyev, R., "A Quarter of a Century with Satellites," *Pravda,* 1982.

22. Anureev, *Antimissile and Space Defense.*

23. Hansen, Lynn M., *Soviet Navy Spetsnaz Operations on the Northern Flank: Implications for Defense of Western Europe* (College Station: Center for Strategic Technology, Texas A&M University System, April 1984).

24. *Soviet Military Power.*

25. Kipp, *Soviet Views on Military Operations in Space.*

26. Zhukov, G., "To Prevent the Militarization of Space," *Izvestiya,* June 2, 1984, p. 5.

27. Kipp, *Soviet Views on Military Operations in Space.*

PART FIVE

SDI and Arms Control: Compatible or Conflictual?

10

Arms Control and Strategic Defense: The Past Need Be No Prelude

Keith B. Payne
Patrick Friel

The Strategic Defense Initiative (SDI) holds the potential to alter the U.S. approach to strategic policy modestly, radically, or not at all. A radical departure from policy is likely to provide more fertile ground for ballistic missile defense (BMD) deployment than is modest change. Indeed, it will be difficult to sustain an argument for very limited BMD protection. However, the proper combination of BMD deployment and arms control could provide an excellent basis for a radical departure from current policy—while minimizing or eliminating many of the problems that SDI critics foresee.

The near-term deployment of limited ballistic missile defense could be compatible with continued reliance on the traditional approach to strategic deterrence based on the threat of nuclear retaliation. In this case, BMD simply would provide a means of protection, in addition to existing means (i.e., hardening, mobility, and a high state of readiness), to enhance the prelaunch survivability of U.S. retaliatory forces. In contrast, if the SDI ultimately leads to the long-term deployment of comprehensive defenses for the urban-industrial United States, then the now-traditional U.S. policy of deterrence based on mutual U.S. and Soviet vulnerability[1] will have been altered radically.

Of course, if the SDI, as a research program, indicates only that urban-industrial defense is likely to remain infeasible and that BMD for retaliatory forces is not cost effective, then it will have little if any effect on policy. The spectrum of potential effect SDI could have on policy is broader than that of virtually any other strategic program—which is why, of course, it is the focus of so much attention and controversy.

Critics of the SDI have complained that discussion of different defensive missions reflects an ambiguity of intent and purpose behind the SDI—as if defensive missions ranging from the protection of retaliatory forces to the defense of urban-industrial centers were mutually exclusive. However, there need be no tension in these goals. This chapter first will discuss the compatibility between near- and far-term defensive missions; present a possible near-term defensive "architecture"; and, examine how arms control and the deployment of BMD could be compatible in the realization of these defensive goals.

Many former senior officials doubt the notion that arms control can be compatible with BMD deployment and the spectrum of potential defensive missions.[2] Such a judgment rests, of course, on a fairly straightforward and familiar line of reasoning.

1. Both sides desire an offensive retaliatory deterrent.
2. That deterrent can be attained readily in the absence of defenses.
3. Consequently, the absence of defense provides a basis for reducing nuclear weapons to the minimum necessary for the purpose of deterrence.
4. Defenses, although unlikely to be effective, will cause both sides to doubt their deterrents and increase their offensive capabilities to counter the defensive threat to their respective retaliatory potential.[3]

However, this presumed incompatibility between arms control success and BMD deployment and the presumed compatibility between the absence of defense and offensive arms control may exist only in the theory underlying the foregoing sentiment. This analysis will suggest why a defensive-oriented strategic environment could provide a more suitable basis for offensive arms control than has the offense-dominated environment of the past. In short, it will suggest why the familiar arms control theory offers less hope than an approach to arms control more compatible with BMD. A BMD system that serves primarily to add uncertainty to Soviet attack planning does not represent a radical departure in the U.S. policy of deterrence—it simply reinforces the current retaliatory threat for deterrent effect.

Ballistic Missile Defense and Deterrence

There are three questions that must be addressed in any consideration of a BMD system that is limited in potential to the reinforcement of retaliatory deterrence through the defense of retaliatory forces: Is the enhanced survivability gained at the launch end lost in penetration as a result of Soviet BMD? Is BMD the most cost-effective means of enhancing prelaunch survivability? Can BMD for retaliatory forces compete cost effectively with Soviet offensive countermeasures? Unless U.S. policy is intended, eventually, to move beyond retaliatory deterrence, these questions will dominate consideration of the strategic role BMD can usefully play in U.S. policy. For example, could launcher mobility and/or superhardening provide a means of enhancing intercontinental ballistic missile (ICBM) prelaunch survivability that would not lead to the additional level of penetration difficulty associated with Soviet BMD deployment? It is not, as yet, clear that answers to such questions indicate a unique role for BMD. Proponents of near-term BMD missions, such as the defense of ICBMs, are likely to find the rationale for very limited BMD deployment difficult to sustain in the face of these considerations. As long as BMD is evaluated in the context of continued reliance on a policy of retaliatory deterrence, the specter of Soviet BMD negating (or more than negating) the enhanced prelaunch survivability provided by U.S. BMD will pose a difficult barrier to breach.

However, if the ultimate goal is to move beyond deterrence through retaliatory nuclear threat, then the rationale for initial BMD deployment, even though of limited capability, will be easier to sustain. In this case, there need be less concern about the prospect of losing strategic targeting access to the Soviet Union. Indeed, it is assumed that both sides will lose targeting access to the other's homeland. If the objective is to create a defensive strategic environment wherein strategic retaliatory threats no longer will play the key role for deterrence, then the "specter" of Soviet BMD will no longer be as compelling.

Nevertheless, the timing of the U.S. move away from a retaliatory nuclear deterrence must be considered carefully. A retaliatory deterrence will continue to be essential, at least during the initial period of BMD deployment, to help sustain stability. It will be important to ensure that the combination of Soviet offensive and defensive capabilities does not introduce a period of significant U.S. disadvantage during the initial transition period. In this regard, a BMD system that can enhance the prelaunch survivability of U.S. forces could be helpful; in effect it would undercut the value of the enormous Soviet counterforce offensive buildup of the last decade. It therefore would help maintain the U.S. retaliatory deterrent during this period wherein it otherwise might degrade prematurely.

However, it is critical that a U.S. BMD system not be limited in its potential to the protection of ICBMs. Rather, it should provide for the defense of retaliatory forces while establishing a defensive layer that can itself be augmented by additional defensive layers in support of a more comprehensive defense objective. This suggests that simple near-term systems, the useful potential of which is limited to the defense of ICBMs, may not have a stand-alone role in the defensive transition. Rather, a near-term architecture with greater potential would be appropriate.

This requirement does not lead to a dead end for BMD. A near-term, midcourse BMD system could support multiple defensive missions, including the protection of selected point targets such as ICBM silos and/or large area targets such as urban-industrial centers. The very large defensive "footprint" of a ground-based, midcourse BMD system would contain numerous military and civilian assets; the system would not have to be dedicated to the defense of a single type of target. Indeed, a significant benefit of ground-based, midcourse BMD is that the Soviet Union could not be certain as to which targets within the enormous defended area would be protected.

This is not to say that any midcourse BMD could provide complete protection for every potential target within a defended area—far from it. Rather, selected assets—whether civilian centers and/or ICBM silos—might be defended preferentially. (Preferential defense raises the probability of successful defense for some candidate aimpoints, while conceding a lower probability for the protection of others.) If the Soviet Union could not be certain, prior to its attack, what the United States would attempt to defend or with how many interceptors, the defense could achieve a significant advantage. The Soviet Union would be compelled to target all aimpoints it intended to attack as if they were defended. These factors introduced by strategic defense would cast additional uncertainties on any Soviet first-strike planning.

Possible Defensive Architecture

This brief examination of the potential relationship between BMD and arms control is not predicated on distant BMD technologies. Indeed, what is described here is a potential BMD system that if complemented by relatively modest offensive force limitations could help stabilize any initial "transition period" to future and more comprehensive defensive capabilities.

A description of this type of midcourse defense system is possible. In broad terms, this technology primarily would involve a constellation of space-based active and passive sensors that would unambiguously identify reentry vehicles (RVs) and provide the defense with a precision impact point prediction of each offensive weapon. The kill mechanism would be a ground-based, infrared homing, nonnuclear impact warhead (with the option of a very small nuclear warhead) that would operate outside the atmosphere or in the high endoatmospheric flight regime.

Two other sensors could be part of this general BMD system concept and would execute the same BMD functions. They would provide the same data as the space-based sensors but much later in the engagement. These sensors are an airborne optical system with a long wavelength infrared (LWIR) sensor and a ground-based, high-frequency, imaging radar, the two key sensors in the third tier of the SDI concept.

The BMD technology suggested in this analysis involves two constellations of satellites, one in geosynchronous orbit (25,000 miles) and a second in low, near-earth orbit at about 600 miles. The task of the geosynchronous satellites would be to track the infrared radiation emitted by the various stages of the launch of a Soviet offensive booster. If the Soviet offensive boosters were of the older, single warhead type, then the twenty-year U.S. experience with early warning satellites suggests that the amount of infrared radiation from the final correction, or vernier, stage could easily be detected from geosynchronous orbit.[4] Because the single RV is separated from the booster at that point, the position and velocity vector of the weapon are known rather precisely. Consequently, it should be possible to predict where the RV will impact in the United States. With this precision impact point information from the satellite system, the BMD battle manager could launch an infrared homing, nonnuclear interceptor from the United States long before the threat appears at the horizon for a sensor located on the ground in the United States. The defense interceptor would acquire and track the RV with its infrared sensor and then deploy an antennalike "net" to which is attached high-density metal spheres.[5] The RV would be destroyed by the enormous energy associated with its impact with the net. The net warhead was successfully tested in the U.S. Army's Homing Overlay Experiment (HOE) in 1984, an experiment in which a high-fidelity nuclear warhead was destroyed outside the atmosphere by a nonnuclear impact warhead. HOE was the only verified RV kill experiment ever conducted by the United States in a BMD field experiment.[6]

Of course, "modern" Soviet ballistic missiles, ICBMs, or submarine-launched ballistic missiles or SLBMs (SS-17, SS-18, SS-19, SSN-8, SS-N-18, SS-N-20, etc.) are MIRVed. To provide the same precision RV impact point prediction data, therefore, the geosynchronous satellite must be able to observe and track the very low infrared radiation associated with the multiple, independently targeted, reentry vehicle (MIRV) bus plume. The detection and tracking of the radiation from the MIRV bus plume are very difficult tasks involving the detection of a radiation source equal to several light bulbs at 25,000 miles.[7] Consequently, in order to track RVs deployed from MIRV buses and provide the required precision impact point predictions, it will be necessary to introduce one of three (and possibly all three) constellations of satellites in lower orbit.

These sensor options include thirty to forty passive LWIR sensors viewing the MIRV bus above the horizon and a comparable number of active sensors—that is, millimeter wave radars or lasers.

These space-based sensors would be deployed in very low earth orbit at about 550 to 625 miles. This orbit is chosen so that the sensor will be below the belt of intense ionizing radiation formed by high-energy charged particles that are trapped in the earth's magnetic field, the so-called Van Allen belt. It is generally accepted that it would be very difficult, if not impossible, to make these relatively sophisticated sensor systems survive and perform their BMD functions in this intense radiation environment.[8]

The task of these low altitude satellites would be to detect and track RVs as they are deployed from the MIRV bus and provide a precision impact point prediction on each to the BMD battle manager. The low-altitude constellation would search the region where the bus is separated from the last stage of offensive booster and thus acquire and track the bus before RVs are deployed. The data from the geosynchronous satellites would be used for this initial bus acquisition function. The required LWIR sensor technology is well within the state of the art.[9] Millimeter wave satellites are not now the state of the art but are a reasonable projection thereof, particularly within the framework of the technology associated with the SDI. The millimeter wave radar would provide a high-quality range-doppler image of the MIRV deployment phase and then obtain a precise track on each. The other imaging sensor option would be an ultraviolet or infrared laser that would acquire a high-quality, two-dimensional image and a range-doppler image of the MIRV deployment phase. The lasers required are beyond the state of the art and with no agile beam capability in the foreseeable future.[10]

Thus, the credible technologies that could be employed in the second, lower constellation during the next decade or so would be the passive LWIR sensors and the millimeter wave radars. The primary task of this two-sensor system would be to obtain high-quality passive and active (range-doppler) images of the MIRV deployment sequence that would allow the defense to reject unambiguously a wide spectrum of offensive confusion devices called penetration aids. Through a detailed observation of the MIRV deployment phase, these new space-based sensors could present an opportunity to solve the key technology issue that has plagued the U.S. BMD program for twenty-five years: the decoy or penetration aids "discrimination" problem. Presumably, an important objective of the SDI program is to demonstrate the ability of these sensors, including lasers, to perform successfully the discrimination task during the MIRV deployment phase in a field demonstration. These sensors also will track each identified RV and provide a precise impact point prediction, particularly the millimeter wave radar (less than 1 mile).

It is instructive to estimate the costs of this BMD concept, assuming deployment of both a passive and an active sensor system in low earth orbit. It could be argued that only one of these two sensor systems would be required, preferably the millimeter wave radars that provide excellent RV impact point predictions very early in the total engagement time and offer greater decoy discrimination capability. However, the passive LWIR sensor system provides yet another phenomenology, in addition to the geosynchronous infrared and the radar data, that many consider essential to verify a Soviet launch of any consequence. The LWIR sensor also would execute another important mission: surveillance and tracking of any space object. In addition, the BMD radar satellite will support another important U.S. intelligence and military mission—namely, ocean surveillance with a high-quality imaging capability.

The unit costs of NASA's sophisticated visible and infrared Landsat imaging satellites is close to $500 million.[11] The annual Department of Defense budget for military space projects indicates that communications satellites cost from $150 to $250 each.[12] Let us assume that the three sophisticated BMD satellites would cost $500 million each and add a 50 percent penalty for survival aids (e.g., hardening, maneuvering, decoys, concealment, etc.) for a total of $750 million each.[13] Assuming eight high-altitude elliptical or geosynchronous satellites plus a nominal deployment of seventy active and passive satellites in low orbit, the total satellite costs would be about $58.5 billion. If the nonnuclear warheads are assumed to cost about $3 million each—a relatively high figure[14]—the total BMD system costs for a 1,000-warhead capability would be about $61.5 billion; a 10,000-warhead capability would cost about $85.5 billion. These cost estimates, however imprecise, are not outlandish in the SDI context and are at least traceable to deployed satellites or a tested BMD component (i.e., the successful HOE nonnuclear warhead demonstrated in 1984).

As we shall see, arms control measures could make the "near-term," nonnuclear kill BMD system described in this chapter attractive. This version of a nonnuclear, nationwide BMD with the appropriate arms control measures would provide a wide spectrum of strategic arms control and technical advantages. Perhaps the most important technical advantage would be that the initial deployments of the geosynchronous satellites with some 620-mile orbit LWIR or millimeter wave radars could literally "practice the war" by tracking Soviet and U.S. ballistic missiles tests, or even Chinese, French, and British missile tests. This would mean that the key technical feature of any BMD system—the ability to acquire and provide precision tracks on actual RVs—would be demonstrated continually in real time using operational U.S. and Soviet ballistic missiles as targets. This is a unique and powerful technical advantage of a BMD that employs space-based sensors. No ground-based or aircraft-based BMD could even remotely achieve a comparable technical capability.

Since the United States initiated its early warning satellite program twenty years ago, thousands of tracks of missiles launchers of all types have been acquired. However, all these tracks were of the missiles themselves and not of RVs. The space-based BMD sensors discussed here would provide a precise count of the numbers of reentry vehicles along with a precision trajectory for the missile, the MIRV bus, and the RVs. While the space based sensors are continuously obtaining tracking data on missile tests, it also should be possible to safely conduct actual engagements, using high-fidelity simulated warheads, on some of the ballistic missile tests using a part of the ground-based nonnuclear interceptor system derived from technology demonstrated in the 1984 HOE experiment.

A review of the history of U.S. BMD research and development during the past twenty-five to thirty years would show that the BMD sensor technology eventually was used to improve dramatically the arms control verification, intelligence gathering, and early warning capability of the United States and the Western world. Two examples are the Cobra Dane radar in Shemya, Alaska, and the shipborne Cobra Judy radar.[15] The radar technology and the associated data processing system were derived extensively from the BMD technology program. Most importantly, the software ("the brains" of both) was derived from the data base obtained by the midcourse and reentry phenomenology program conducted on U.S. ballistic missile tests at the Kwajalein missile test range in the Pacific during the past twenty years.[16] These sensors have provided an invaluable data base on Soviet missile tests into Kamchatka, including the characteristics and number of MIRV buses and RVs. These data on ballistic missile tests are obviously of enormous value in verifying key provisions of SALT I and SALT II (e.g., the number of warheads tested on each missile). The space-based sensors associated with the nationwide BMD system discussed here would be an even more dramatic improvement in the Western world's verification capability.

The low-altitude constellations would provide a unique, powerful verification tool for arms control. These sensors would be used mainly to verify tests at the principal U.S. and U.S.S.R. missile test ranges: Kamchatka for the Soviet Union and the Kwajalein missile test range in the Pacific for the United States. In addition, these space-based BMD sensors would provide an absolutely unique (and presently nonexistent) capability to monitor missile tests of various kinds, including penetration aids development tests, in the two inland ranges (White Sands Missile Range in New Mexico for the United States and the Sary Shagan BMD test site for the USSR). One of these three space-based base sensors, the millimeter wave radar, also may be able to monitor tests of other strategic weapons: for example, cruise missiles launched from a variety of platforms.[17] In addition, these BMD sensors also could contribute to other vital strategic missions (e.g., early warning and general intelligence gathering). Therefore, even a modest deployment of these BMD sensors (meaning the ability to acquire and precisely track RVs) would increase enormously the ability of both the United States and USSR to monitor and verify the development of a wide spectrum of strategic weapon systems.

The development and eventual deployment of even a modest version of these three space based BMD sensors (i.e., the geosynchronous short wavelength infrared satellites as well as the millimeter radar and long wavelength infrared satellites in low orbits) surely represents a "no lose" opportunity for the United States and the Western world. Their development and deployment, therefore, is certainly a worthy near-term objective of President Ronald Reagan's SDI.

A combination of qualitative and quantitative restrictions on offensive ballistic missiles on both sides would have an even greater impact on the strategic value of this version of a nationwide BMD. The capability of this BMD is limited by the ability of the satellite to track RVs and the kill vehicle inventory. The tracking capability of the low-altitude LWIR satellites with modern, on-board data processing could be enormous (perhaps thousands of RVs). The tracking ability of the millimeter wave radars is limited by the fact that it has a limited (1–2) electronic scan capability,[18] these are predominately mechanical scan radars with a limited phased-array capability. However, in the framework of the aggressive technology associated with the SDI, it is reasonable to postulate a large increase in the power and phase-array capability of millimeter wave radars in the near term. Thus, the RV tracking ability of the low-altitude millimeter wave radar constellation, with the associated on-board data processing, also could be enormous.

The strategic capability of this BMD system, therefore, would be controlled entirely by the threat and kill vehicle inventory. The system's effectiveness could be increased significantly and gradually after the initial space-based sensors and the limited kill vehicle inventory have been deployed. Thus, a systematic decrease in the offensive warhead inventory coupled with an increase in the nonnuclear kill vehicle inventory of this BMD system might facilitate a gradual, *controllable* transition from the present "assured destruction" strategic nuclear regime to any "assured survival" regime.

Arms Control

Arms control, if pursued properly, could facilitate a stable defensive transition in several important ways. First, it could regulate defensive deployment so that neither side would fear that the other was acquiring an extremely asymmetrical BMD potential. Second, it could reduce the offensive threat for each party, thereby allaying fears that the opponent (as seen from Washington or Moscow) could acquire a potent "sword *and* shield."[19] Third, it could hasten the attainment of a significant damage limitation capability by both sides and reduce the offensive potential of each. In short, arms control could provide an avenue to a "stable transition"; it could ameliorate the "destabilizing" potential of strategic defense suggested by critics of the SDI.[20]

The arms control regime that could provide such benefits would have to include both offensive and defense constraints, although these constraints would be quite different. Defensive limitations would facilitate the gradual deployment of ground-based BMD systems, such as the midcourse system described previously or the Soviet Galosh system for Moscow and/or ABM-X-3 systems. Current testing and deployment restrictions of the ABM Treaty on "exotic" and space-based interceptor components could be eased gradually as the offensive potential of both sides' arsenals declined. This would help ensure the effectiveness of ground-based defenses and allay fears that one side would gain a space-based "shield" while retaining a "sword."

Offensive limitations and reductions could facilitate the effectiveness of both sides' defenses in a controlled fashion so as to prohibit either side from acquiring a significantly advantageous combined offensive/defensive capability. Offensive limitations would be most helpful if they combined quantitative and qualitative limitations. For example, the gradual reduction of ICBM, submarine-launched ballistic missile (SLBM), and antisatellite (ASAT) launchers, a MIRV ceiling such as that incorporated in SALT II, and prohibition on the testing and deployment of advanced ballistic missile penetration aids, new cruise missiles, and strategic bombers would help reduce anxieties on both sides concerning the transition away from strategic offensive forces. Such offensive limitations and reductions would facilitate the effectiveness of both sides' defenses and provide a controlled, predictable, (at least) partially cooperative transition.

To critics of the SDI, this type of arms control regime, which stresses offensive restrictions and greater defensive freedom, appears farfetched. The traditional view of arms control is that the absence of BMD facilitates offensive arms control, while the deployment of BMD prohibits any prospect for the control of offensive forces. SALT I, for example, was premised on such considerations.[21] However, it is argued here that this traditional approach to arms control has proven to be a failure on its own terms; an arms control regime that reduces offensive forces while endorsing and facilitating the effectiveness of strategic defenses has a relatively greater chance for success.

Ambassador Gerard Smith placed the question in the proper perspective in recent testimony before the Senate Committee on Arms Services.

> The preamble to the [ABM] Treaty states in part, "proceeding from the premise that the limitation of antiballistic missiles systems, as well as certain agreed measures with respect to the limitation of strategic offensive arms, would contribute to the creation of more favorable conditions for further negotiations on limiting strategic arms." The President's March 23, 1983 program, on the contrary, is based on the premise that the decontrol of antiballistic systems will lead not only to limits on strategic arms but also the elimination of nuclear weapons. Which of these two premises is closer to the truth? That is the basic question which faces this Administration and this Congress and will face their successors.[22]

Although the complete "elimination of nuclear weapons" is utopian from any perspective, there are clear reasons why a defense-oriented approach to arms control should hold more potential than has the traditional offense-dominated approach advocated by Ambassador Smith. The Reagan administration has not yet presented these reasons convincingly—that does not mean that the case cannot or should not be made.

The traditional approach to arms control has proved to be such a failure for the same reason that defense-reliant arms control should be more successful. The Soviet objective of acquiring a strategic damage limitation potential destroyed the logic of traditional arms control and led to its failure in SALT. That same Soviet objective provides the basis for the success of defense-reliant arms control. These assertions can be supported sequentially.

First, SALT I and II reflect the implementation of the traditional approach to arms control. At the time of SALT I it was expected that the ABM Treaty and SALT I had stopped "[Soviet] momentum in the offensive weapons area" and, in particular, frozen the number of Soviet heavy ICBMs.[23] The Soviet Union long since had achieved a retaliatory deterrent capability; thus, in the absence of BMD, offensive limitations would be realized. Indeed, Ambassador Smith stated at the time of SALT I that there would be a "commitment on their part [i.e., the Soviets] not to build any more of these ICBMs that have concerned us over the years" and that, based on U.S. and Soviet desire, reductions in heavy ICBMs would be forthcoming within two or five years (i.e., in SALT II).[24]

The primary impetus behind the U.S. drive to freeze and then reduce the Soviet heavy ICBM potential was to reduce the Soviet counterforce threat to U.S. retaliatory forces—notably ICBMs[25]—and thereby support the offense-reliant notion of deterrence and stability. Virtually the entire edifice of SALT was built on this rationale and reflected the traditional approach to arms control.

Of course, history has moved in the opposite direction. Soviet heavy ICBMs, according to the U.S. SALT I definition, were neither frozen nor reduced; rather, following SALT I they doubled in number and increased 1,600 percent in warheads. As a result, U.S. retaliatory forces, especially ICBMs, have become increasingly vulnerable.

Obviously, the Soviet Union did not consider acceptable an arms control regime that would enforce a reduction in its offensive counterforce potential. Yet the success of SALT, from the U.S. perspective, depended on the Soviet Union giving up its apparent attachment to improved counterforce offensive capabilities. The Soviet Union chose not to reduce its offensive counterforce capabilities; it also chose to increase those same capabilities significantly, largely in circumvention of the U.S. understanding of SALT I.[26]

It appears that the Soviet Union pursues a combined arms program in support of a damage limitation potential. Soviet counterforce offensive capabilities play an important role in that program. Consequently, the Soviet Union has been unwilling to abide by the traditional U.S. approach to arms control. Stephen Meyer recently observed that "first and foremost, Soviet military doctrine holds that the primary mission of the Soviet armed forces is to prevent the nuclear devastation of the Soviet homeland—i.e., the armed forces must be able to deter strategic nuclear war, and yet be able to limit damage should deterrence fail."[27]

This same damage limitation objective that undercut the traditional approach to arms control and the U.S. agenda in SALT could provide the motivation necessary for Soviet acceptance of a defense-reliant approach to arms control. The traditional U.S. approach to arms control, to be successful, required the Soviet Union to comply with U.S. mutual vulnerability notions of strategic objectives and stability and to deny Soviet damage limitation capability. The Soviet Union chose not to comply with U.S. views and pursued its combined counterforce/active defense program for damage limitation. However, defense-reliant arms control would acknowledge and accept the Soviet damage-limitation objective; indeed, it would facilitate the Soviet achievement of that objective. Such an arms control regime would grant the Soviet Union what it long has sought and what has been denied it by the U.S. offensive force programs—a significant damage limitation capability. This is a major advantage of defense-reliant arms control. It recognizes Soviet objectives and the Soviet qua Soviet perspective on strategic policy. It establishes an arms control regime that would not require the Soviet Union to do violence to its own interpretation of its self-interest, as does the traditional U.S. approach to arms control.

The U.S. SALT agenda would have required the Soviet Union to abandon its damage limitation objective by limiting both counterforce offense (in the form of heavy ICBM restrictions) and BMD. SALT, in effect, sought to mandate a Soviet policy compatible with U.S. concepts of stability through mutual vulnerability. Rather than abandoning its damage limitation objectives, the Soviet Union simply placed heavy reliance on counterforce offense following SALT I, and the U.S. purpose in SALT, as presented to Congress, collapsed. Whereas traditional arms control sought to compel Soviet policy away from damage limitation, defense-reliant arms control would structure an arms control regime on the basis of Soviet and U.S. damage limitation goals. Significant damage limitation capabilities may be impossible without such an agreement—hence, the possible motive for both sides to agree to a defense-reliant regime.

There is evidence that at least some Soviet military authorities already have become dissatisfied with continued pursuit of damage limitation through counterforce offense.[28] Defense-reliant arms control would provide an alternative; it would grant the Soviet Union (and the United States) its damage limitation objective through strategic defense.

The "cost" to both the United States and the Soviet Union of such an arms control regime would be in terms of offensive potential. Both superpowers in effect would be forced to modify their offensive targeting plans in return for acquiring a damage limitation capability. The United States has already indicated its acceptance of the concept of offensive capability reduction as the price for strategic defense.[29] Because Soviet offensive targeting policy serves largely to support an ultimate damage limitation mission, the Soviet Union should find defense-reliant arms control attractive. Defense-reliant arms control, in effect, would force the Soviet Union to choose between a capability to ensure "dead Americans or live Russians."[30] Under the traditional approach to arms control the Soviets sought offensive and defensive capabilities. Defense-reliant arms control, for the first time, would present the Soviet Union with a choice: Continue with the current offense-dominant course and damage limitation will continue to elude you, or participate cooperatively in offensive limitations and defensive deployment and acquire that long sought after capability. A defensive approach to arms control has this inherent motivational factor (a large "carrot") that traditional arms control rejects. An additional "carrot" for the Soviet Union, of course, is that U.S. participation in such an arms control regime would help ensure that the United States could not acquire a "sword and a shield." It would allay any Soviet fears that the SDI and strategic defense are simply vehicles for a U.S. bid to regain strategic superiority.

In short, traditional arms control requires the Soviet Union to do violence to its own view of strategic objectives; when implemented in SALT, that approach to arms control largely failed because the Soviet Union proved unwilling to endorse U.S. concepts of strategic policy and stability. In contrast, defense-reliant arms control recognizes Soviet strategic objectives, provides a framework for Soviet achievement of the primary goal of damage limitation, and does so in a way that is compatible with U.S. strategic objectives. Defense-reliant arms control, logically, should be a more effective basis for achieving offensive reductions. Indeed, it would provide meaning to even modest offensive limitations that otherwise have little but political value. Modest limitations, such as those of SALT I and II, had little value in an offense-dominant strategic environment in terms of reducing the effective level of damage that nuclear war would cause and had no utility in "capping," much less reducing, the accumulation of "destabilizing" countersilo weapons. Yet even modest offensive limitations, in a defensive environment, could prove meaningful by enhancing the effectiveness of defensive, damage limitation systems.

As has been elaborated elsewhere, a defense-reliant approach to arms control has significant advantages from a verification standpoint.[31] For example, in an offense-dominant environment it would be difficult for the United States to accept deep offensive reductions with a party that is known to engage in noncompliance.[32] Modest levels of cheating may not be "militarily significant" when there are very high offensive force levels. Yet in the absence of defenses, even modest levels of cheating could be quite significant at low offensive force levels. Strategic defense would act as a safeguard against "significant" cheating in a deeply cut offensive force level. Thus, it would be impossible to disarm superpower offenses without a transitional substitution of defensive for offensive capabilities. As this assessment illustrates, strategic defense appears to be essential for genuine offensive reductions.

The Soviet Union and Defense-reliant Arms Control

A critical question follows from this discussion: If a defense-reliant approach to arms control would provide the Soviet Union with what it seeks but cannot obtain through traditional arms control in an offense-dominated strategic environment, why has the USSR rejected the notion of limiting strategic offensive forces while loosening restrictions on strategic defenses? In short, why is the Soviet Union so apparently opposed to the SDI? Why is the Soviet Union *not* behaving in the manner suggested in this analysis?

It probably is true that the Soviet Union considers the SDI to be politically fragile and is attempting to use arms control as a means of influencing U.S. policy concerning the SDI. The Soviet Union historically has used arms control to affect the U.S. domestic political debate.[33] The Soviet Union can increase political pressure against the SDI by announcing that arms control and strategic defense are incompatible but that arms control will happen if the SDI is curtailed. The Soviet Union almost certainly preferred the pre-SDI strategic environment wherein it alone pursued serious strategic defense programs, had the initiative in BMD development and deployment, and pursued offensive countersilo capabilities. The Soviet Union held the initiative in countersilo offensive and defensive programs. The SDI at least threatens to counter the enormous Soviet investment in countersilo ICBMs and to take the defensive initiative out of Soviet hands and place it in a high-tech arena where the United States probably holds substantial long-term advantages. It is not difficult to imagine that the Soviet Union would prefer a strategic environment where the United States is not competing defensively against Soviet counterforce capabilities or with the Soviet BMD R&D program. Consequently, if the Soviet Union can help stop the SDI by influencing the U.S. political process, it will have moved the clock back to a more acceptable context with a minimum of resources expended. To encourage defensive arms control, the United States must demonstrate to the Soviet Union that "moving the clock back" in such a fashion is not an option.

The Soviet Union is unlikely to endorse defense-reliant arms control or to say anything sympathetic about U.S. defensive objectives until it is clear that the United States is committed to the program. Until it is clear that the SDI cannot be undone through intervention in the U.S. political process, the Soviet Union probably will try to influence that process to stop the program.

However, once it is clear that a choice can and must be made between counterforce offense and active defense, the Soviet Union is likely to choose the latter as a more feasible avenue to damage limitation. Obviously, there are no certainties in such considerations. However, what is certain is that the traditional offense-reliant approach to arms control has failed, on its own terms, in a two decade, "real-world" test. It failed because it did not take the Soviet danger limitation objective into adequate consideration. A defense-reliant approach to arms control, such as is outlined here, addresses that inadequacy in traditional U.S. arms control policy and offers an alternative that should provide a more effective basis for arms control.

In summary, this analysis has attempted to discuss the SDI and arms control in realistic terms sure to offend almost everyone. It is not clear at this point that near-term BMD of very limited potential would provide benefits that would outweigh the costs, if the United States is to continue its reliance on retaliatory strategic threats for deterrence. Of course, if the Soviet Union "breaks out" of the ABM Treaty with rapid BMD deployment, those "costs" would be inflicted regardless of U.S. actions. However, barring such a development, it appears that the rationale for BMD of very limited potential will be difficult to sustain. However, a midcourse system of ground-based interceptors could provide near-term protection for retaliatory forces and the basis for expansion into the thicker BMD "architecture" necessary for more comprehensive, urban-industrial defense. In this case, the U.S. policy course would have moved away from reliance on mutual vulnerability for stability; and a gradual loss of targeting access to the Soviet Union would be anticipated.

Arms control could play a critical, perhaps essential, role in a "defensive transition." It could help regulate that transition, make it more predictable, and contribute significantly to the effectiveness of U.S. and Soviet defenses. Such a "cooperative" transition would minimize or eliminate most of the potential instabilities suggested by SDI critics. The immediate objective of the combination of near-term defensive technology and offensive arms control would not be to eliminate the policy of retaliatory nuclear deterrence. Rather, the objective would be to enhance deterrence by complicating any Soviet first-strike planning.

The notion of arms control aiding and abetting strategic defense is absolutely foreign to traditional U.S. thinking about arms control. However, that traditional approach to arms control has proven itself a failure and a very poor model upon which to base expectation. A defense-reliant approach to arms control addresses the weakness of the traditional approach and could facilitate a stable transition to a much more defense-oriented strategic environment.

Notes

1. The logic of this deterrence policy is most recently and clearly presented in Thomas Schelling, "What Went Wrong With Arms Control," *Foreign Affairs* 64, no. 2 (Winter 1985/86):219–233.

2. See McGeorge Bundy et al., "The President's Choice: Star Wars on Arms Control," *Foreign Affairs* 63, no. 2 (Winter 1984/85):264–278.

3. This chain of logic concerning the alleged relationship between strategic defense and arms control was presented repeatedly during the ABM debate of 1968–1972. See, for example, Abram Chayes et al., "ABM Deployment: What's Wrong With It?" in Abraham Chayes and Jerome Wiesner, eds., *ABM—An Evaluation of the Decision to Deploy an Antiballistic Missile System* (New York: Signet Books, 1969), pp. 3–60.

4. The assumption made here is that vernier engine radiation, although much smaller than the radiation emitted by space-shuttlelike boosters (megawatts), is in turn much larger than the few "light bulbs" emitted by the MIRV bus.

5. For a description of the two successful nonnuclear intercepts of simulated ballistic missile warheads in the exosphere, the Homing Overlay Experiments, see Ed Vaughan, *U.S. Army Ballistic Missile Defense Systems Command* (Huntsville, Ala.: Public Information Office), with animated film cassette. Also see description of the intercept of a Minuteman ICBM RV on June 10, 1984 in a test over the North Pacific. This event, the product of an earlier R&D program, has been subsumed under the president's Strategic Defense Initiative (SDI). See International Institute for Strategic Studies (IISS), *The Military Balance, 1984–1985* (London: International Institute for Strategic Studies, 1984), p. 3.

6. All the previous BMD "intercepts" were in fact experiments in which the BMD interceptor (Nike Zeus, Spartan, or Sprint) passed close enough to the incoming reentry vehicle so that a detonation of the interceptor warhead would destroy the RV. See Benson D. Adams, *Ballistic Missile Defense* (New York: Elsevier, 1971), p. 55. The HOE was the first physical destruction in the exosphere of an RV containing a high-fidelity, nuclear warhead simulation.

7. The ability of space-based sensors to track during the bus deployment or postboost phase is one of the "critical technologies" mentioned in the U.S. Department of Defense, *Strategic Defense Initiative, Defensive Technologies Study* (Washington, D.C.: GPO, April 1984), p. 12.

8. The Van Allen belt is a region of intense ionizing radiation in space about the earth formed by high-energy, charged particles that are trapped by the earth's geomagnetic field. It was discovered by NASA's Explorer satellites in the late 1950s and early 1960s and is named after the experiment's sponsor, Dr. Van Allen of NASA/AMES Research Center.

9. The LWIR sensor parameters were estimated by assuming that a typical RV in space remained close to room temperature and emitted a few tens of watts of radiation at 10 microns. This amount of radiation should be detectable at 500 to 1,000 miles with 3 to 7 foot collecting aperture using cooled, doped, silicon detector focal plane. The radar parameters were obtained by assuming that typical RVs have radar cross-sections significantly less than one square meter and that a relatively high detection probability is desired at 1–1,250 miles with millimeter wave power sources limited to 5 kilowatts, so that a 49 to 66 foot collecting aperture would be required.

10. For one of the few public discussions on the state of the art in high-power laser technology, see W. D. Henderson, "Space Based Lasers," *Aeronautics and Astronautics* (May 1982).

11. The annual budget for Landsat during the four- to five-year procurement phase was $80–90 million so that a $500 million total program cost per satellite is not unreasonable. See "Fiscal 1983 Aerospace Budget," *Aviation Week and Space Technology* (February 15, 1982):24. The requested annual budget for the surveillance, acquisition, and kill assessment elements of the SDI program for fiscal year 1986 was $1.4 billion, which is presumably for the development and not the procurement of these complicated space-based sensors. See "Fiscal 1986 Aerospace Budget," *Aviation Week and Space Technology* (February 11, 1985):19. The annual air force strategic surveillance and warning budget is $500–600 million, which probably includes an annual procurements budget of not more than one early warning satellite per year. See "Fiscal 1983 and 1984 Aerospace Budgets," *Aviation Week and Space Technology* (February 15, 1982):17 and (February 7, 1983):19.

12. The costs of the Defense Satellite Communications System from the annual Department of Defense budget for fiscal year 1985 is $291.2 million for the procurement of two spacecraft. See "Fiscal 1985 Aerospace Budget," *Aviation Week and Space Technology* (February 6, 1984):19. The NASA 30/20 GHZ satellite is apparently much more expensive as indicated in "Fiscal 1984 Aerospace Budget," *Aviation Week and Space Technology* (February 7, 1983):26. The program costs for this new high technology satellite were estimated at $350 million.

13. The 50 percent cost penalty is an estimate that assumes a complete survival aid package: nuclear hardening, a maneuvering motor, decoys, electronic and optical countermeasures, signature reduction.

14. The smaller Patriot Area Defense Missile costs about $1.7 million each according to the "Fiscal 1985 Aerospace Budget," p. 17. (The budget for 585 Patriot missiles was $1.096 billion.) A cost estimate of $3 million for the more sophisticated homing, nonnuclear BMD missile is therefore not unreasonable.

15. The Cobra Dane radar at Shemya, Alaska, and the shipborne Cobra Judy radars are described in IISS, *The Military Balance, 1985–1986* (London: International Institute for Strategic Studies, 1985), p. 7.

16. The data acquisition on strategic missiles by the Kwajalein Atoll radars have been a matter of public record for more than a decade.

17. The space-based, millimeter wave radar can acquire targets near the surface of the earth in any kind of weather. Visible and infrared sensors cannot obtain any data near the earth's surface under bad weather conditions. The space shuttle's shuttle imaging radar (SIR) dramatically demonstrated the power of space-based imaging radars to obtain images of the earth's surface. The Public Information Office of NASA's Jet Propulsion Laboratory, Pasadena, California, will provide a collection of radar images of the earth's surface from the SIR experiment on the space shuttle in 1981.

18. A full two-dimensional, phased-array, millimeter wave radar would involve more than 100 million elements, well beyond the state of the art demonstrated in the Cobra Dane and Cobra Judy radars.

19. The Soviet Union has indicated that it has such a fear. See, for example, *"Star Wars": Delusions and Danger* (Moscow: Military Publishing House, 1985).

20. See, for example, Hans Bethe et al., "Space-Based Ballistic Missile Defense," *Scientific American*, 25, no. 4 (October 1984); also see the statement of Ambassador Gerard Smith before the Senate Committee on Foreign Relations, *Nuclear Strategy, BMD and Arms Control*, November 12, 1985, p. 3.

21. See, for example, Paul Nitze, "SDI and the ABM Treaty," *Current Policy*, no. 711 (Washington, D.C.: U.S. Department of State, Bureau of Public Affairs, June 1985), p. 1; and Paul Nitze, "The Objectives of Arms Control," *Survival* 27, no. 3 (May/June 1985):98–107.

22. See the statement of Ambassador Gerard Smith before the Senate Committee on Armed Services, November 21, 1985, p. 6. The same question is posed somewhat differently in U.S. Congress, Office of Technology Assessment, *Ballistic Missile Defense Technologies*, OTA-ISC-254 (Washington, D.C.: GPO, September 1985), p. 33.

23. See statements by Melvin Laird and Henry Kissinger respectively in U.S. Senate, Committee on Armed Forces, *Military Implications of the Treaty on the Limitations of Anti-Ballistic Missile Systems and the Interim Agreement on Limitation of Strategic Offensive Arms*, 92nd Cong., 2nd sess., pp. 121, 545. (Hereafter, *Military Implications*.)

24. See Gerard Smith, in *Military Implications*, pp. 100, 296, 395; and Nitze, "The Objectives of Arms Control," pp. 102–103.

25. See portions of Henry Kissinger's SALT I testimony as reprinted in *SALT I Reconsidered* (Washington, D.C.: Institute of American Relations, 1979), pp. 5, 34, 77, 85, 86, 100; see Gerard Smith's testimony reprinted in the same document, p. 37; and that of Melvin Laird, pp. 81–82.

26. For a discussion of how the Soviet Union circumvented the U.S. effort to achieve meaningful heavy ICBM restrictions in SALT I, see Keith B. Payne, *Strategic Defense: "Star Wars" in Perspective* (Lanham, Md.: Hamilton Press, 1986), pp. 164–165.

27. Stephen M. Meyer, "Soviet Strategic Programmes and the U.S. SDI," *Survival* 27, no. 6 (November/December 1985):278.

28. See *Krasnia Zvezda*, May 9, 1984, translated in "Orgarkov on Military Technology," *Survival* 26, no. 4 (July/August 1984):187–188; and M. A. Gareev, *M. V. Frunze—Voennyi Teoretik* (Moscow: Voenizdat, 1985), excerpted and translated in Leon Gouré and Michael Deane, "The Soviet Strategic View," *Strategic Review* (Fall 1985):102–106.

29. See Nitze, "SDI and the ABM Treaty," pp. 2–3.

30. To use the late Donald Brennan's apt phase in "The Case for Population Defense," in Johan Holst and William Schneider, Jr., eds., *Why ABM?* (New York: Pergamon Press, 1969), p. 116.

31. See Keith B. Payne and Colin S. Gray, "Nuclear Policy and the Defensive Transition," *Foreign Affairs* 62, no. 4 (Spring 1984):820–842.

32. Colin S. Gray, "Moscow Is Cheating," *Foreign Policy*, no. 56 (Fall 1984):141–152.

33. See Keith B. Payne and Dan Strode, "The Soviet Approach to Arms Control and Its Implications," *Soviet Union/Union Sovietique* 10, pts. 1–3 (1983):218–243.

11

Strategic Defense and Arms Control: Are They Compatible?

Stephen J. Cimbala

This chapter suggests that U.S. initiatives to develop and deploy active strategic defenses, including ballistic missile defense (BMD), are not necessarily inconsistent with U.S.-Soviet nuclear arms control. Much expert opinion is skeptical of the compatibility of the Reagan Strategic Defense Initiative (SDI) and arms control. The prognosis for any U.S. theater or strategic BMD may be more pessimistic to the extent that it is considered incompatible with arms control agreements that improve strategic stability.

Stability in the U.S.-Soviet strategic balance is not a static concept. The balance of forces and the doctrines that make stability possible are subject to change. This is true for crisis stability, or avoidance of the temptation to preemption, and arms race stability. There is no reason in principle why defenses cannot contribute to stability along with appropriate arms control agreements. Indeed, efforts to preserve stability on the basis of past concepts and doctrines might be counterproductive. If the superpowers were to deploy technologies that create partial rather than total strategic defenses, as seems likely, their effects on stability and their compatibility with arms control would be complex.

Arms control agreements are not judged here as ends in themselves. However desirable they might be in the abstract, policymakers will evaluate them according to other criteria. U.S. presidents and their advisers will be concerned with the effects of strategic defenses and arms control on the credibility of U.S. deterrence and on the survivability of valuable assets if deterrence fails. The effects of defenses on the superpowers' perceptions of one another's intentions and capabilities are also important and as yet undetermined. The U.S. and Soviet leaders of the future may be required to grope toward solutions that stabilize offensive and defensive mixes within the constraints of tolerable costs and perceptions of threat. Their tasks will not be enviable and certainly not determined by any formula concepts of strategy.

The future environment of technology may be permissive of defense-dominant arms agreements. As important as technology is the military doctrine (in the Soviet sense of that term) that the two superpowers have, which informs their military deployments and war plans. What wars are they likely to have to fight and about what? There is in the United States a large hiatus between the routine planning for war and the development of political contexts by which wars are judged meaningful or not. However one might dispute the Soviet formulation of these things, that they have sought to address explicitly the connection between war and politics, including nuclear war application, must be noted. If we are serious about arms control in the West, we are obligated to think through the matter of what to do if deterrence fails, beyond destroying as much of the Soviet Union in retaliation as we have got the ordnance to do. Arms control reminds us that there are individual and societal values on the receiving ends of those warheads. Efforts to protect those values if deterrence fails may be counterproductive if those efforts make war more likely, but this cannot be decided in the abstract. The probability and costs of war are interdependent. If past practice is any guide, the United States first will deploy those technologies that seem workable and then decide what to do with them.

Vulnerability

Most discussion of the vulnerability of U.S. and Soviet societies or forces to nuclear attack have focused on the equivalence, or lack thereof, of megatonnage, reentry vehicles (RVs), throw weight, and other gross indicators of the military balance. Pre- and postattack ratios based on these enumerations are certainly an important part of the question of strategic stability.[1] But they are far from sufficient to characterize it fully.

Postattack exchange ratios become important insofar as their anticipated values influence the calculations of the aggressor and defender before war begins.[2] The objective of policy is deterrence in preference to war. There is substantial evidence that the Soviets, despite disclaimers to the contrary, are quite sensitive to the importance of these calculations of expected postexchange utilities.[3] However, the methodology of U.S. planners can be described in ways that are quite unflattering.[4] Force exchange models do have limitations (that have been acknowledged by their users).[5]

Whether U.S. and Soviet vulnerabilities are essentially equivalent is only one important issue, however. Equally important is the absolute as well as the relative level of those vulnerabilities.[6] If both superpowers expect that their societies will be subject to destruction in the postattack period, regardless of what they do to one another's forces in the initial exchanges, they will have every incentive to "overcompensate" in retaliatory forces for the protection that they cannot provide to their societies. This overcompensation is the dynamic that drives the arms race, to the extent that it is dependent upon action-reaction dynamics at all. Because societies cannot be defended, deterrence cannot be permitted to fail. However, as the possibility that deterrence may fail cannot be discounted, one must purchase extra insurance against that possibility. This extra insurance can be threatening to the opponent, depending on the quantity of insurance and the opponent's perceived liability.

Counterforce first-strike capabilities are an example of asymmetrical action-response. They are unnecessary to maintain strategic stability, but they are attractive offsets in the minds of policy planners to societal vulnerability. If one cannot prevent the destruction of one's cities and economic values before they are attacked, then the only option is to destroy the forces of the opponent before they are used. Close students of Soviet planning have noted Soviet interest in preemption and in prompt, hard-target capabilities.[7] This interest is often lamented in U.S. discussions. But the Soviet perspective is not illogical at the level of war plans, even if Soviet political leaders would prefer not to emphasize it in their declaratory policy. Indeed, it is their only choice if they hope to limit significantly the damage inflicted on their society once war begins.

The United States is in exactly the same position with regard to its employment policies. Although declaratory policies from the mid-1960s until 1974 emphasized mutual assured destruction (MAD) and countersocietal retaliation, actual war plans differed from this benchmark.[8] U.S. strategic targeting since the early 1960s has been primarily countermilitary rather than countersocietal; additions to the Single Integrated Operational Plan (SIOP) since then have been primarily designed to provide more discrete options for attacking military and command and control targets.[9]

Although they have not fully succeeded in doing so, successive U.S. administrations have emphasized in their policy statements the desire to extend deterrence into war itself and to terminate nuclear war at the lowest possible level of violence consistent with the attainment of U.S. policy objectives and the denial of Soviet objectives.[10] This attempt to establish policy guidelines for the waging of nuclear war with discrimination among policy objectives and levels of destruction has been subjected to much unfair criticism. As Albert Wohlstetter has noted, in a recent, important article, there is a strong case for the technological and political plausibility of war limitation rather than for Armageddon.[11] Where mutual vulnerability theory has gone wrong is in its attachment to absolutist models of nuclear war outcomes; policymakers want forces that are controllable and options that are realistic.[12]

Notice that this contention, that policymakers prefer wars that do not get out of hand, does not mean that history always will validate their efforts. World War I provides ample illustration of the process by which crises can slide into wars. Richard Ned Lebow's exhaustive study, *Between Peace and War,* notes the variety of ways in which crisis stability can be demolished by misperception and erroneous expectations on the part of decisionmakers.[13] Richard Smoke has described the various processes of escalation and historical cases in which the process of escalation proved less controllable than leaders had anticipated.[14] Charles Hermann has studied political crisis behavior from many perspectives, and in a recent paper reviews unilateral U.S. and bilateral U.S.-Soviet crisis avoidance proposals. His conclusion is that both the United States and the USSR have "gradually engaged in a variety of activities that have seriously eroded the stability of their deterrence systems to withstand the effects of a direct international crisis without ending in war."[15]

Skepticism about the ability of policy planners to prevent war or limit its effects should deterrence fail can be found among those who have studied extensively the U.S. system for strategic command, control, and communication (C^3). Critics have noted that the U.S. strategic C^3 system is based on physical facilities that are subject to rapid destruction; that the president and the secretary of defense (collectively known as the National Command Authority, or NCA) may not survive the initial exchanges of nuclear weapons; that their successors may not be knowledgeable about war plans even if they were located; that communications among sensors, force commanders, and NCA can be jammed, degraded by electromagnetic pulse (EMP), or otherwise rendered inoperable; and, that those U.S. strategic forces for which the launch vehicles are most survivable (submarine-launched ballistic missiles, or SLBM) are the ones that have the most serious problems in maintaining communications with national command authorities after war begins.[16]

The more that evidence accumulates about the probable effectiveness of the U.S. strategic C^3 system, the more implausible it has seemed that nuclear war could be fought in a controlled manner for a protracted period. Thus, students of the command and control system have joined analysts who focus on decisionmaking in concluding that controlled nuclear warfighting is a wrongheaded policy prescription. They may be right. But their pessimism rests on at least one important, but frequently implicit, assumption: that damage cannot be limited to the United States to any significant degree once deterrence fails. It is the awareness of this vulnerability, and the undoubted Soviet awareness of the USSR's own substantial vulnerability (whether equivalent or not), that makes crises unstable. Mistakes cannot be accommodated because the consequences of those mistakes cannot be redeemed.

The reaction to this state of affairs by many analysts has been to deplore any movement away from mutual vulnerability. This resistance to change is understandable. First, the mutual vulnerability of the two sides to substantial destruction undoubtedly has helped to prevent some direct clashes between their conventional or nuclear forces. Even those who doubt that the USSR considers societal vulnerability important concede that the U.S. ability to threaten nuclear retaliation has proved useful on more than one occasion.[17] Second, movement away from the present condition of U.S. declaratory and employment policy toward a more defense-oriented posture may not be fully controllable by U.S. planners. To some extent, U.S. reactions to Soviet development and deployment of new defensive and offensive systems will be part of the calculus of transition.[18] Third, defense dominance may not be attainable. It may turn out, given the uncertainties of technology, that improved defenses may be superseded by even more improved offenses. A stable plateau of nonnuclear defenses dominating nuclear offenses might not be possible, even if it were deemed desirable.[19]

These concerns are valid. They must be addressed by those who want to make the case for greater defense emphasis in U.S. strategic declaratory, employment, and acquisition policies.[20] The next section of this chapter attempts to do just that and to place the issues of defense emphasis or defense dominance in a somewhat different, less polarized context.

Defense of Offense Dominance?

From a strategic standpoint, the pursuit of defense dominance as anything more than a temporary respite from a permanent competition is illusory. Historical case studies, including U.S.-Soviet strategic arms deployment, show that offenses are sometimes temporarily dominant, defenses another. This last observation may seem trite, but it is not. Offensive and defensive technologies as well as the skill with which military planners and political leaders apply them determine the strategic environment in any given war or era.[21] Thus, we need to identify those conditions that usually but temporarily favor the offense and those that favor the defense in war. Some qualifications of what is typical in this regard, of course, will be necessary if these general categories are to be applied to nuclear war between superpower adversaries. The implications for arms control will then be drawn out.

Several factors favor the offense or offensively oriented strategies in war. The most important of these is the advantage of surprise and initiative. The side striking first has the advantage of possibly catching the other off guard. In the nuclear age, being caught off guard could be catastrophic, depending upon the kind of attack being made and the relative success of the attacker.

Analysts concerned about nuclear surprise have focused on the possibility of either a "bolt-from-the-blue" Soviet attack against U.S. forces on day-to-day alert or an attack that achieves tactical if not strategic surprise against "generated" forces.[22] According to the U.S. Department of Defense, preemption is the strategy preferred by Soviet military planners, but their forces also are prepared for launch under attack or retaliation following U.S. attacks.[23] Surprise might be an important "force multiplier" for the USSR if it were attacking preemptively, depending upon U.S. readiness and doctrine. U.S. strategic forces have been sized and doctrine predicated on worst-case assumptions about surprise as a baseline.[24]

The importance of surprise in a superpower nuclear exchange should not be exaggerated, however. Strategic surprise (in which the intentions of the attacker are completely concealed until the last moment) is improbable given modern surveillance and intelligence collection methods. A far more plausible scenario would be the decision paralysis of U.S. policymakers faced with ambiguous but plausible warnings they refused to believe.[25] Tactical surprise might be obtained by the USSR in terms of the method and timing of attack. The Soviets have options for attacking U.S. strategic nuclear forces, conventional forces, command centers and supporting connectivity, war supporting industry and resources, communications, and power supplies. Either selective or omnivorous attacks would be possible with current and near-term Soviet forces. There has been some evidence in recent years that Soviet doctrine is more sympathetic to limited nuclear options in the event of war, although that evidence as understood by Western scholars is not undisputed.[26]

Successful tactical surprise might make some difference for the Soviets in the relative survivability of U.S. land-based missiles and bombers. Immediate postexchange residuals would favor the Soviets in prompt, hard-target capabilities compared to U.S. capabilities surviving the Soviet attack. U.S. retaliation against remaining Soviet forces (given present forces on both sides) would not improve the asymmetry in prompt, hard-target capabilities if the Soviets (as expected) launched a third strike.[27] Nor would the United States improve its societal recovery position relative to that of the USSR by retaliating.[28]

But it is important to note that such relative comparisons of postattack destruction can be misleading if the absolute levels of destruction on both sides exceed historical precedent, which they easily could. Moreover, policymakers on either side might not know with any assurance exactly how much destruction their forces or society had absorbed. Vulnerability of the command system complicates the problem of postattack damage assessment. Preattack expectation that postattack damage assessment could be precluded makes fine comparisons between indicators seem irrelevant to real war plans.[29] If political leaders do not expect that high-fidelity attack assessments will survive the transattack period, and there is every reason to be pessimistic, then relative advantages in a war of absolute ferocity will seem unimportant. The flexibility and endurance of the command system thus interact paradoxically with the importance of surprise. Surprise becomes less important the more improbable the survivability of the command system (of the opponent) and its fidelity appear to be.

It might seem counterintuitive that surprise is less important if the opponent's command system is more vulnerable. The statement is wrong if the opponent's command system can be destroyed totally, thereby limiting retaliation to what can be readily absorbed by the attacker. But that outcome is not the most probable one in the U.S.-Soviet strategic exchange case. Instead, the more vulnerable command system will disintegrate only partially. The attacker by surprising its adversary will receive uncoordinated retaliation from the defender's isolated but still functioning primary and secondary commands. Thus, a partial destruction of the U.S. command system by Soviet attackers could leave them worse off. They would have to deal with a decentralized rather than a centralized command infrastructure on the opposite side, thereby precluding war termination on any terms at all.[30] It has been asserted that the Soviets are disinterested in war termination once nuclear war begins and that they would fight any such war without reservation or limit. But this expectation about Soviet priorities is as dependent upon what the United States does as on what the USSR does. A U.S. strategic command system that is rigged on the principle of top leadership acting as a "safety catch" holding back retaliation from primary and secondary force commanders, as Bracken has explained it, penalizes any attacker who would destroy partially but not totally the C^3 system.[31]

If surprise favors the attacker when all other things are considered equal, then "friction" favors the defender. Friction is the collective expression for the multitude of things that can go wrong with war plans and political expectations.[32] Friction has both a deliberate and an autonomous quality. Some reversals in fortune or failures in plans are the result of enemy action. Enemy intelligence may anticipate your intentions or enemy commanders may outfight yours. On the level of autonomous reasons, plans are difficult to implement. It is sometimes the case, as Bruce G. Blair has noted with regard to the U.S. strategic command system, that the relationship between top leaders' intentions and ground-level implementation is coincidental.[33] This does not happen because middle and lower levels in organizations are mischievous by definition but because those levels have their own priorities, perceptions, and expectations. Organizations are hierarchical on paper but in practice depend upon tacit bargains among units. Getting the work of an organization done on a day-to-day basis may require cooperative interactions contrary to explicit leadership norms and policies.[34]

Friction could defeat a well-planned strategic or tactical nuclear surprise. The sources of friction in preemptive nuclear attack would be numerous. They are both technical and political. Technical factors include all the things that might go wrong with launch orders and trajectories, guidance mechanisms, unexpected gravitational anomalies, wind conditions, and other parameters.[35] Political uncertainties include the possibility of responses by the defender other than those anticipated by the attacker. Even partial defenses for the United States, for example, would add uncertainty to Soviet preemptive attack calculations. Comparable Soviet systems would complicate U.S. estimates about the penetrativity of U.S. warheads toward their assigned targets.

Other kinds of friction could take over after war had passed the transattack stage of the initial, almost automatic exchanges of strategic weapons.[36] Political leaders could lose their nerve, be killed, or abdicate. Popular dissidence and revolution against established authority in a postattack environment are not inconceivable, especially to leaders in the Kremlin who will have studied World War I with great intensity. Attempts to stop a war in progress by the side that was "winning" the counterforce war might be impossible if command destruction proceeded too "successfully." A real dilemma for U.S. planners is uncertainty about Soviet expectations for protracted war. Whereas it appears that U.S. declaratory policy emphasizes protracted war because it might be limited in its intensity, Soviet military doctrine has conceived of protracted nuclear war as global and almost impossible to limit.[37]

Some other factors that might favor superpower nuclear attackers or defenders should be noted. Offensive strategies will be more successful in the hands of those who profess and attempt to structure their forces for preemptive attack. There are some claims that this is precisely what the USSR has done, but the claims are disputed.[38] Protracted war preparedness might favor the defender if the war were prolonged for weeks or months. The side with the more survivable, enduring command system also would be more suited for a defensive role. Thus, it seems paradoxical that the USSR, which has allegedly developed a more robust "control structure" compared to the United States, seems to have a more offensively oriented "strategic culture." Technical factors may account for some of this because Soviet strategic forces are apparently maintained on lower levels of day-to-day alert than are their U.S. counterparts. Political factors also may help to explain the apparent paradox. The Soviet control structure is necessary to keep the pervasive control of the *nomenklatura* from being challenged.[39] Still, the paradox may point to something important to U.S. planners. The USSR is a more formidable adversary on the defensive than it is when waging offensive war from its borders, if assessments of its conventional war-fighting capabilities are any indicator.[40] It might be preferable for Soviet strategists to put more of their strategic eggs into survivable, second-strike weapons systems and into slow, as opposed to prompt, counterforce. Soviet deployments of mobile ICBMs, such as the apparently imminent SS-X-24, and the development and deployment of new and more capable land- and sea-based cruise missiles could point to movement of forces structure in this direction.[41]

Friction could bedevil war plans if those plans were based on excessive optimism about capabilities. In a U.S.-Soviet nuclear exchange, unrestricted countercity exchanges could destroy not only both societies, but also affect significantly the international climate and ecology.[42] Yet no policy planner or academic expert provides optimistic reassurance that war can be controlled. Extensive academic studies and journalistic research both point to the same conclusion about the U.S. strategic C^3 system: Its capabilities for limiting strategic exchanges once the first few tens of weapons have been launched are not impressive, even if the Soviets choose to "cooperate."[42] Air force officials have acknowledged that the SIOP can only be fine tuned so far; more options can be added, but more decision time cannot be made available to the president or his (her) successor. As Colin Gray has pointed out, military leaders who fear the opponent's escalation are not going to dissipate their strategic forces on sub-SIOP skirmishing of "limited nuclear options" and finely calibrated responses.[43]

If surprise has tended to favor the offense in war and friction might be said, with some reservations, to favor the defense (insofar as it affects the attacker's expectations about preemption), what can be expected from deployments of new defenses and the offenses to overcome them? Some speculation is unavoidable in any assessment of future prospects but necessary if we are to assess the probable relationship between strategic defenses and arms control. The conditional nature of "defense dominance," as explained in the preceding section, carries important implications for the stability of deterrence and the possibility of arms control with robust defenses.

Strategic Impasse and Arms Control

Deployed strategic defenses by either superpower will not necessarily make arms control easier to attain. Freeman Dyson has outlined the most hopeful alternative for arms control following the deployment of nonnuclear strategic defenses. He calls this the "live and let live" posture. Nonnuclear, space-based defenses would gradually supersede nuclear offensive forces, which would be reduced through negotiation.[44] He contrasts this scenario to the "arms controllers' future" in which no defenses are deployed beyond those permitted by the ABM Treaty of 1972. Dyson argues that strategic defenses could assume benign roles of surveillance and monitoring in support of arms control agreements, in addition to defending against reduced offensive arsenals.[45]

It might turn out that way. But it might also turn out that the problem is more complicated than such a solution will admit. These complications have to do, again, with the distinction between absolute and relative levels of destruction in nuclear war between the superpowers and the importance of those expectations in the calculations of political and military leaders.

Strategic defenses of uncertain effectiveness relative to the offensive forces that oppose them are as likely to cause war as they are to prevent it. It might be thought otherwise by those who make slide rule calculations of counterforce exchanges and postattack residuals. After all, defenses could shift the balance of forces from one favoring the attacker (facing no defenses) to one favoring the defender (if the defenses complicated or defeated the attack plan). The defenses could do this if all that mattered in the estimates of political leaders and military planners were surviving forces. But societies are also at stake. Postattack forces superiority is little consolation for either U.S. or Soviet leaders if their society were reduced to medieval proportions and postattack recovery were precluded by the unprecedented devastation certain to follow from present superpower arsenals.[46]

This situation might change if either the United States or the USSR could deploy defenses that precluded societal destruction. This is not impossible, but it is certainly improbable in the lifetimes of this writer and his readers. The reasons for this improbability are quite clear, if often poorly understood by specialists: The relative value of undestroyed U.S. society exceeds that of a destroyed Soviet society or destroyed Soviet forces under any system of calculation acceptable to the U.S.

people. There could be no such thing as "victory" if U.S. society were reduced to primitive conditions, regardless of the relative success of U.S. "revenge."

Defenses for U.S. society will be hard to come by. They cannot be provided without the cooperation of the Soviet Union. Soviet leaders who are determined to maintain their strategic retaliatory power will invest whatever they have in order to assure the success of that retaliation. To do otherwise is tantamount to surrender in terms of their ideology. Their capacity to destroy U.S. forces in preemptive attacks or retaliatory strikes will vary with the balance of counterforce capabilities and active and passive defenses. But their capability to destroy U.S. society in retaliation has absolute, rather than relative, value for them. This point should not be misunderstood. The Soviet are interested in counterforce capabilities and favor counterforce strategies to the extent that they can gain relative advantage on these dimensions. But their "last resort" threat against U.S. preemption is the same as the U.S. last resort against a Soviet first strike. The United States cannot survive a Soviet retaliation as a society—that is, with the U.S. social identity intact. Both the American Revolution and the October Revolution are the real hostages in this relationship, not the numbers of potentially surviving weapons.

Arms Control

The coupling of U.S. strategic defenses with incentive for U.S.-Soviet arms control still is to be discussed. Such an arms control regime may not be "defense dominant." But it might be "defense catalytic."

Improved defenses that provide partial protection for strategic forces, command and control, and societal values important for war recovery could be part of a defense-catalytic transition to improved strategic stability. The goal is defined this way deliberately. Fewer or less devastating arms are significant only cosmetically unless they improve stability in the short and long run. The short run part of a defense-catalytic transition undoubtedly will use strategic defenses to improve the survivability of forces and commanders, but this is insufficient for the longer term.

Mutual vulnerability theorists postulate a world in which the populations of the United States, Europe, and Russia, plus those that are inadvertent hostages, will accept this fate indefinitely. Rather than accept this fate indefinitely, those made hostage on purpose are likely to rebel, at least in the West. As Michael Howard has noted, the arrival of U.S.-Soviet strategic parity makes the provision of deterrence to the Soviets and reassurance to the Europeans through the U.S. extended deterrence umbrella more difficult.[47] Thus, the longer term transition will have to provide some defense for the populations of the West, if not for the Soviets and their allies as well. This is viewed as problematical for stability by some advocates of mutual assured destruction.

Seen from a larger perspective, however, mutual vulnerability may be adjusting its boundary conditions for mutual stability rather than going out of existence as a result of strategic defenses. Phased in gradually, societal strategic defenses can contribute to stability. They can do so by removing hostage populations from their condition in equivalent stages for both superpowers and their blocs. What might prevent this favorable employment of phased defenses, however, would be a mathematical fetishism that insisted upon identical rather than essentially equivalent phased deployments.

The most significant objection to "graduated societal defense" as suggested here is that if both sides had it, it might make conventional deterrence more important. This cannot be precluded, but if such a situation occurs, it cannot be blamed on the removal of mutual societal vulnerability. Conventional deterrence became more important the day the USSR achieved strategic parity with U.S. strategic forces. SALT II was an official codification of that parity. The imminent countdown on the credibility of U.S. extended deterrence for Europe, in all but the most drastic scenarios, had begun in the early 1980s. Speaking in Brussels in 1979, former secretary of state Henry A. Kissinger told his European audience that the United States should not make threats that it did not mean, and, he added, if it did mean those threats (to defend Europe with nuclear weapons in response to Soviet conventional attack), it should not execute them.[48]

Flexible response is not totally without deterrent value; the absence of war in Europe attests at least in part to the credibility of flexible response in Soviet estimates. But flexible response requires amendment and refurbishing to prevent the erosion of credibility. Two presumably attractive options "on paper" are deficient for want of resources or political will. The first is the restoration of U.S. strategic superiority, with its spillover benefit of extended deterrence as a free rider. The second is robust NATO conventional deterrence, sufficient to defeat in its own terms any Soviet conventional aggression in Europe.[49] Although these two positions have important advocates on both sides of the Atlantic, the consensus of political pragmatists is that they are unattainable.[50]

Active defenses for U.S. and European forces and territory could provide attainable options for improving flexible response. For this to happen, those defenses must be coupled to arms control proposals that are credible to the Soviet Union. Only voluntary restraint by the USSR embodied in bilateral arms agreements with its Western counterpart would allow for the offensive arms limitations prerequisite to meaningful defense. Appealing to the Soviet sense of altruism or attempting to cast the Soviets as international curmudgeons are both unproductive approaches toward this end.

The USSR will not be talked out of its interest in counterforce. Nor should it have to be.[51] U.S. employment policy on counterforce targeting has not been too dissimilar from what we know about Soviet war plans.[52] Several administrations have insisted that the USSR dismantle or drastically reduce the quantity of its ICBM warheads and throw weight because these are counterforce first-strike weapons. But they are counterforce first-strike weapons by U.S. choosing as well. The United States has chosen to deploy its modern land-based missiles in vulnerable fixed silos while U.S. leaders have complained about potential Soviet first-strike capabilities. The MX-Peacekeeper ICBM has received so many reviews of its basing mode that the Scowcroft Commission saw no alternative to "analysis paralysis" except imminent silo basing.[53]

What makes Soviet ICBMs really threatening, however, is not their capability to disarm the Minuteman force. That is their intermediate objective. Their real clout comes from the fact that if they disarmed a significant portion of the U.S. strategic "triad," the USSR might then coerce the United States into surrendering. It is significant that both hawks and doves in the U.S. strategic debate agree on how the Soviets might coerce the United States to prevent retaliation (and both admit that such a strategy is risky). They might coerce the United States because of the residual and ultimate threat to destroy U.S. society if U.S. forces retaliate following their counterforce first strike. In other words, the case for ICBM vulnerability is an assessment of the superstructure, in good Marxist terminology; the substructure on which the effectiveness of the strategy really turns is the threat of societal destruction.

It is to the credit of mutual vulnerability theorists that they clearly perceive this relationship between force vulnerability as a way station to the threat of societal destruction. Counterforce exchange models often confuse means with ends on this score. But mutual vulnerability theorists err by opposing any and all strategic defenses for population, economy, and society. Those defenses, deployed gradually and mutually between the superpowers, may improve strategic stability and arms race stability. If both hawks and doves in the U.S. debate are to be believed, defenses could hardly make things worse.

The qualification "hardly" must now be explained. *Ceteris paribus* conditions do not always apply. The graduated, mutual deployment of strategic defense for societal protection could be destabilizing under certain conditions. Those conditions are perceptions by either or both superpowers that their strategic forces are vulnerable to surprise attack.

Societal protection cannot be stabilizing if the protection of retaliatory forces is neglected. Vulnerable forces on either side contribute to crisis instability. Force survivability is a means to the end of a broader survivability for forces and society together.

Defenses would contribute to crisis instability if they were so weak compared to offenses that they invited preemptive attack. When I say "compared to offenses," I imply the marginal utility of additional investment in defense compared to the utility of offsetting that increment in defense by additional offenses. This criterion of cost effectiveness at the margin has been stated by Paul H. Nitze, U.S. special representative to the superpower arms negotiations at Geneva, as a requirement that any U.S. Strategic Defense Initiative system must meet, among others.[54] Thus, U.S. defenses might be less technologically "sweet" than offenses (or Soviet defenses) but still worth the price if their marginal utilities were cheaper than Soviet offensive "offsets."

It should also be anticipated, although it is not in standard scenarios, that future offenses may not resemble those now deployed by either superpower. The first quarter of the twenty-first century could witness startling developments in "smart" or even "intelligent" strategic systems; electronic brainpower could become more important than the brawn-based indicators to which we are so accustomed. The "military balance" could turn appreciably within that time period on developments in computers, electronics, and directed-energy weapons. One illustration of this indeterminacy in moving from blunt to smart weapons involves the capability of ballistic missiles. Intercontinental missiles of the future might have smart "front ends" that resemble the cruise missiles of today, and they might be fired from land, sea, or airborne launchers in sufficient numbers to saturate defenses bases on present-generation technology. Strategic war with conventional weapons would confound arms control efforts with new and destabilizing challenges. Another illustration of the transition lies in the movement of both U.S. and Soviet military establishments toward mobility and concealment for land-based missiles, both strategic and intermediate.[55]

Conclusions

Superpower strategic defenses can be contributory to arms control. Future arms control agreements may not be based on a defense-dominant world, but they are unlikely to be predicated upon defensive celibacy. Short-term stability can be improved by using defenses deployed gradually and mutually to protect retaliatory forces, commanders, and connectivity. Long-term stability can be improved by providing offsets for the enriched offensive forces of future decades.

Mutual vulnerability theorists and mutual survival advocates may discover mutual respect for one another's positions. This discovery may wait upon the departure of the Reagan administration, given the president's rhetorical commitment to defensive perfection.[56] Totally invulnerable societies leave deterrence with no rationale. Totally vulnerable societies tempt deterrence failure. Partially protected societies with survivable forces provide incentives to keep deterrence working. Societal protection that does not undermine strategic stability should be compatible with arms control agreements that are in the interests of both superpowers. Soviet interest in protecting Soviet society from the effects of war should not be considered pernicious but historically grounded in Soviet experience and politically sensible given that nation's ideology. U.S. interest in societal protection lies in mitigating the effects of war should deterrence fail and in adding credibility to the policy story for extended deterrence. These superpower interests are not identical, but they are convergent if the above arguments are convincing. Past perceptions that U.S. and Soviet perspectives on the defense-arms control connection are hopelessly divergent have prevented agreement.

Notes

The author gratefully acknowledges comments on earlier drafts by Michael Altfeld, Albert Carnesale, and Richard Ned Lebow. Views are solely those of the author.

1. For illustrative assessments, see Matthew Bunn and Kosta Tsipis, "The Uncertainties of Preemptive Nuclear Attack," *Scientific American* 249, no. 5 (November 1983):38–47; and Keith B. Payne, *Nuclear Deterrence in U.S.-Soviet Relations* (Boulder, Colo.: Westview Press, 1982), p. 173.

2. T. K. Jones and W. Scott Thompson, "Central War and Civil Defense," *Orbis* 22, no. 3 (Fall 1978):681–713, esp. pp. 682–684.

3. Joseph D. Douglass, Jr. and Amoretta M. Hoeber, *Soviet Strategy for Nuclear War* (Stanford, Calif.: Hoover Institution Press, 1979), pp. 93–96, discuss qualitative and quantitative aspects of the Soviet estimated "correlation of forces." For a Soviet perspective on U.S. estimates, see Henry Trofimenko, "The 'Theology' of Strategy," *Orbis* 21, no. 3 (Fall 1977):497–516, esp. his comments on estimates by Paul Nitze, p. 503.

4. See Fritz W. Ermarth, "Contrasts in American and Soviet Strategic Thought," in Derek Leebaert, ed., *Soviet Military Thinking* (London: Allen and Unwin, 1981), pp. 50–69. Discussion of the technical uncertainties in targeting strategic forces is provided in Donald M. Snow, "Strategic Uncertainty and Nuclear Deterrence," *Naval War College Review* (November-December 1981):27–41.

5. For an overview of Soviet military thinking on strategic issues, see Raymond L. Garthoff, "Mutual Deterrence, Parity and Strategic Arms Limitation in Soviet Policy," in Leebaert, *Soviet Military Thinking,* pp. 92–124; and Benjamin S. Lambeth, "How to Think about Soviet Military Doctrine," in Douglas J. Murray and Paul R. Viotti, eds., *The Defense Policies of Nations* (Baltimore, Md.: Johns Hopkins University Press, 1982), pp. 146–153. It appears that Soviet planners do careful calculations of strategic force balances, pre- and postattack, but also consider qualitative factors in a more inclusive framework than is typical of Western assessments.

6. See Robert Jervis, "Why Nuclear Superiority Doesn't Matter," *Political Science Quarterly* 94, no. 4 (Winter 1979-80):617–633.

7. Douglass, *Soviet Strategy for Nuclear War,* p. 18.

8. Desmond Ball, "Counterforce Targeting: How New? How Viable?" *Arms Control Today* 11, no. 2 (February 1981), reprinted with revisions in John F. Reichart and Steven R. Sturm, eds., *American Defense Policy* (Baltimore, Md.: Johns Hopkins University Press, 1982), pp. 227–234.

9. See, for example, Aaron L. Friedberg, "The Evolution of U.S. strategic 'Doctrine'—1945 to 1981," in Samuel P. Huntington, ed., *The Strategic Imperative: New Policies for National Security* (Cambridge, Mass.: Ballinger Publishing, 1983), pp. 53–99; and Lynn Etheridge Davis, "Limited Nuclear Options: Deterrence and the New American Doctrine," in Christoph Bertram, ed., *Strategic Deterrence in a Changing Environment* (Montclair, N.J.: Allenheld, Osmun and Company, 1981), pp. 42–62. Friedberg suggests that the United States never has had a consensual strategic doctrine if one includes war plans as well as declaratory policies.

10. Leon Sloss and Marc Dean Millot, "U.S. Nuclear Strategy in Evolution," *Strategic Review* 12, no. 1 (Winter 1984):19–28.

11. Albert Wohlstetter, "Between an Unfree World and None," *Foreign Affairs* 63, no. 5 (Summer 1985):962–994.

12. For an exception, see Robert J. Art, "Between Assured Destruction and Nuclear Victory: The Case for the 'MAD-plus' Posture," *Ethics* 95, no. 3 (April 1985):497–516.

13. Richard Ned Lebow, *Between Peace and War: The Nature of International Crisis* (Baltimore, Md.: Johns Hopkins University Press, 1981), esp. pp. 148–228.

14. Richard Smoke, *War: Controlling Escalation* (Cambridge, Mass.: Harvard University Press, 1977).

15. Charles F. Hermann, "The Ultimate Crisis of the Nuclear Era," in Stephen J. Cimbala, ed., *Challenges to Deterrence in the 1990s* (New York: Praeger, forthcoming, 1987).

16. See, for example, Bruce G. Blair, *Strategic Command and Control: Redefining the Nuclear Threat* (Washington, D.C.: Brookings Institution, 1985); Paul Bracken, *The Command and Control of Nuclear Forces* (New Haven, Conn.: Yale University Press, 1983); and Desmond Ball, *Can Nuclear War Be Controlled?* Adelphi Papers, no. 169 (London: International Institute for Strategic Studies, Autumn 1981).

17. As Thomas Schelling has put it, the ability to threaten "just plain war" between superpowers is a formidable deterrent. On the Cuban missile crisis,

see Graham T. Allison, *Essence of Decision: Explaining the Cuban Missile Crisis* (Boston: Little, Brown, 1971); Schelling's discussion of the Cuban missile crisis appears in his *Arms and Influence* (New Haven, Conn.: Yale University Press, 1966), pp. 95–96, and is especially instructive on the difference between wars that begin for foreseeable causes and those that result from miscalculations and conflict spirals.

18. Keith B. Payne and Colin S. Gray, "Nuclear Policy and the Defensive Transition," *Foreign Affairs* 62, no. 4 (Spring 1984):820–842.

19. This is the prognosis, among others, of the Union of Concerned Scientists, *The Fallacy of Star Wars* (New York: Random House, 1984). The authors' contention is that space-based defenses cannot work against present-generation offenses using appropriate countermeasures. The greater challenge for the SDI will be its effectiveness against offenses based on new technologies or new applications of existing technologies.

20. For a clear discussion of the distinctions among these types of policies, see Donald M. Snow, "Levels of Strategy and American Strategic Nuclear Policy," *Air University Review* 35, no. 1 (November-December 1983):63–73.

21. This distinction between available technology and the skill of its application is illustrated in the excellent study by Martin Van Creveld, *Supplying War: Logistics from Wallenstein to Patton* (London: Cambridge University Press, 1977).

22. Illustrations of the differences in U.S. force survivability under generated versus day-to-day conditions can be found in Congress of the United States, Congressional Budget Office, *Modernizing U.S. Strategic Offensive Forces: The Administration's Program and Alternatives* (Washington, D.C.: GPO, May 1983).

23. U.S. Department of Defense, *Soviet Military Power: 1985* (Washington, D.C.: GPO, April 1985), p. 28. Emphasis in Defense Department assessments has shifted from Soviet doctrinal interest in preemption or launch under attack to Soviet preparedness for protracted war, implying force and command survivability.

24. Richard K. Betts, *Surprise Attack: Lessons for Defense Planning* (Washington, D.C.: Brookings Institution, 1982), p. 229.

25. Political authorities might be overwhelmed and confused by large amounts of ambiguous information during a nuclear crisis; warning and intelligence systems might dump too much, rather than too little, on them. See Bracken, *The Command and Control of Nuclear Forces,* pp. 58–59. Soviet interest in the advantages of surprise might lead the USSR to preemption in anticipation of U.S. preparations for attack. For conflicting estimates, see Robert P. Berman and John C. Baker, *Soviet Strategic Forces: Responses and Requirements* (Washington, D.C.: Brookings Institution, 1982), p. 36; and Ermath, "Contrasts in American and Soviet Strategic Thought," p. 66.

26. See Wohlstetter, "Between an Unfree World and None," pp. 983–986, for a summary discussion of recent research on this point.

27. Congress of the United States, Congressional Budget Office, *Counterforce Issues for the U.S. Strategic Nuclear Forces* (Washington, D.C.: GPO, January 1978).

28. Jones, "Central War and Civil Defense."

29. Problems of assessment are described in Bracken, *The Command and Control of Nuclear Forces;* and in Peter Pringle and William Arkin, *SIOP: The Secret U.S. Plan for Nuclear War* (New York: Norton, 1983), pp. 97–100.

30. Bracken, ibid., pp. 117–128.

31. As Bracken explains, "What makes a command-structure attack so difficult is the fact that the Soviets do not know precisely who has nuclear authority in an emergency and, more important, how this authority would flow at the highest levels of government in a crisis." (Ibid., p. 236.) But the physical survival of command assets is uncertain at best. According to Blair, "The obliteration of the ground portion of the U.S. C^3I structure is a Soviet prerogative." (Ibid., p. 187.) Thus, transattack positive control depends entirely upon the airborne command network.

32. Karl von Clausewitz, *On War,* edited and translated by Michael Howard and Peter Paret (Princeton, N.J.: Princeton University Press, 1976). For an insightful discussion of friction applied to modern air power and air strategy, see Barry D. Watts, *The Foundations of U.S. Air Doctrine: The Problem of Friction in War* (Maxwell AFB, Ala.: Air University Press, December 1984).

33. Blair, *Strategic Command and Control,* pp. 65–78.

34. Problem solving under conditions of "bounded rationality" has been established as typical organizational behavior in authoritative studies; see James March and Herbert Simon, *Organizations* (New York: John Wiley and Sons, 1958).

35. These are discussed in Bunn, "The Uncertainties of Preemptive Nuclear Attack."

36. Comments by Brent Scowcroft, *Strategic Nuclear Weapons, Policies and the C Connection* (Bedford, Mass.: Electronic Systems Division, USAF and MITRE Corporation, October 13–14, 1981), p. 94.

37. I am grateful to Robert Jervis for emphasizing this point in correspondence to me on November 20, 1984. Although the Soviets no longer contend that war between capitalism and (Soviet) socialism is inevitable, or inevitably nuclear if deterrence fails, they are not optimistic that U.S.-Soviet strategic war can be limited. On this point see Ball, *Can Nuclear War Be Controlled?;* and Douglass, *Soviet Strategy for Nuclear War,* p. 8. For Soviet views, see V. Zemskov, "Characteristic Fea.ures of Modern Wars and Possible Methods of Conducting Them," in *Selectd Readings from MILITARY THOUGHT, 1963–1973,* selected and compiled by Joseph D. Douglass, Jr. and Amoretta M. Hoeber (Washington, D.C.: GPO, 1974), pp. 48–56.

38. Some sources claim the Soviets are wired only for preemption. See the interview with Lowell Wood in William J. Broad, "Star Warriors," *Science Digest* (September 1985):36–39, 84–85. Soviet preemption might be met by U.S. launch under attack, which has never been precluded by U.S. officials. See Richard L. Garwin, "Launch Under Attack to Redress Minuteman Vulnerability," *International Security* 4, no. 3 (Winter 1979-80):117–139; and John Steinbruner, "Launch Under Attack," *Scientific American* (January 1984):37–47. Garwin offers a technologically enhanced launch under attack option; Steinbruner ex-

presses skepticism that the United States could rely on it in plausible attack scenarios.

39. Harriet Fast and William F. Scott, *The Soviet Control Structure: Capabilities for Wartime Survival* (New York: National Strategy Information Center, 1983); and Richard Pipes, *Survival Is Not Enough: Soviet Realities and America's Future* (New York: Simon and Schuster, 1984).

40. This point is emphasized in Edward N. Luttwak, *The Grand Strategy of the Soviet Union* (New York: St. Martin's Press, 1983), pp. 18–19. Contemporary Soviet conventional offensive capabilities may provide new options. See James M. McConnell, "Shifts in Soviet Views on the Proper Focus of Military Development," *World Politics* 37, no. 3 (April 1985):317–343.

41. Whether mobile ICBMs would provide slow or prompt counterforce would depend upon their tasking by Soviet planners. On the SS-X-24, see U.S. Department of Defense, *Soviet Military Power: 1985,* p. 30.

42. Richard P. Turco et al., "The Climatic Effects of Nuclear War," *Scientific American* (August 1984):33–43; and Anne Ehrlich, "Nuclear Winter: A Forecast of the Climatic and Biological Effects of Nuclear War," *Bulletin of the Atomic Scientists* (April 1984), supplement, 3S-14S.

43. Colin S. Gray, *Nuclear Strategy and Strategic Planning* (Philadelphia, Pa.: Foreign Policy Research Institute, 1984), p. 23.

44. Freeman Dyson, *Weapons and Hope* (New York: Harper and Row, 1984), pp. 272–285.

45. Ibid., p. 280. BMD might not be necessary if substantial reductions in offensive forces could be achieved without it, according to Dyson. But defenses could play an important catalytic role in getting from low levels of offensive forces to zero.

46. Soviet societal vulnerabilities are significant if we are talking about U.S.-USSR SIOP-RISOP exchanges. See Gray L. Guertner, "Strategic Vulnerability of a Multinational State: Deterring the Soviet Union," *Political Science Quarterly* 96, no. 2 (Summer 1981):209–223; John Weinstein, "All Features Grate and Stall: Soviet Strategic Vulnerabilities and the Future of Deterrence," in Robert Kennedy and John Weinstein, eds., *The Defense of the West: Strategic and European Security Issues Reappraised* (Boulder, Colo.: Westview Press, 1984), pp. 39–76; and U.S. Arms Control and Disarmament Agency, *An Analysis of Civil Defense in Nuclear War* (Washington, D.C.: ACDA, December 1978).

47. Michael Howard, "Reassurance and Deterrence: Western Defense in the 1980s," *Foreign Affairs* (Winter 1982-83):309–324.

48. Henry A. Kissinger, "NATO: The Next Thirty Years," in Christoph Bertram, ed., *Strategic Deterrence in a Changing Environment* (Montclair, N.J.: Allanheld, Osmun and Co., 1981), p. 109.

49. Earl C. Ravenal, "Europe Without America: The Erosion of NATO," *Foreign Affairs* 63, no. 5 (Summer 1985):1020–1031.

50. John J. Mearsheimer, "Nuclear Weapons and Deterrence in Europe," *International Security* 9, no. 3 (Winter 1984/85):19–46. Mearsheimer argues that NATO will have to struggle in order to prevent asymmetries in the NATO–Warsaw Pact force balance from worsening in the decades ahead. His argument is in the context of objection to "no first use" proposals.

51. Dyson, *Weapons and Hope*, pp. 250–253, makes valuable distinctions between Soviet counterforce concepts and U.S. limited nuclear war doctrines. He argues that Soviet doctrine is more realistic given Soviet history, although both concepts are (according to Dyson) crisis unstable.

52. Differences in Soviet and U.S. conceptual approaches to war and deterrence might make little difference once central strategic war escalated beyond its initial stages. Doctrines of limited nuclear warfighting and escalation control on the U.S. side would fall short of the technology needed to implement them. Soviet counterforce attacks still would leave the Soviets with postattack forces sufficient to destroy most major U.S. cities, transportation networks, refineries, chemical industries, and other critical assets. Differences in U.S. and Soviet targeting are assumed to follow from differences in acquisition and deployment policies. The United States relies more heavily on SLBMs and the USSR on ICBMs for retaliatory power. These differences might be important in the early stages of war; transattack exchanges would emphasize prompt, hard-target destruction of missile silos and perhaps command centers. Only if a war could be terminated at the end of countersilo exchanges, however, would the Soviet prompt counterforce emphasis provide meaningful dividends. See Congress of the United States, Congressional Budget Office, *Retaliatory Issues for the U.S. Strategic Nuclear Forces* (Washington, D.C.: GPO, June 1978), esp. pp. 14–15.

53. President's Commission on Strategic Forces (Scowcroft Commission), *Report* (Washington, D.C.: April 1983).

54. Paul H. Nitze, "On the Road to a More Stable Peace," in U.S. Department of State, *Current Policy*, no. 657 (Washington, D.C.: February 20, 1985).

55. See Carl H. Builder, *Strategic Conflict Without Nuclear Weapons* (Santa Monica, Calif.: Rand Corporation, April 1983), for a discussion of strategic nonnuclear war. The U.S. rationale for mobile land-based missiles is provided in Scowcroft Commission, *Report*. Also see Jonathan Medalia, *Small, Single-Warhead Intercontinental Ballistic Missiles: Hardware, Issues and Policy Choices* (Washington, D.C.: Congressional Research Service, Library of Congress, May 26, 1983). This report is periodically updated.

56. The White House, *The President's Strategic Defense Initiative* (Washington, D.C.: January 1985), includes a presidential "Foreword" that defines the objective of SDI as "aimed at eventually eliminating the threat posed by nuclear armed ballistic missiles" (p. i). The strategic concept articulated by Nitze expresses similar ambitions, although providing more emphasis on the role of arms control ("On the Road to a More Stable Peace"). Administration scientists concede that no "perfect" defense is possible, but they clearly anticipate that future BMD technology can trump offensive technology.

About the Contributors

Stephen J. Cimbala teaches political science at Pennsylvania State University and has contributed to the field of national security studies for many years, including research on strategic and European nuclear forces, arms control, and U.S. defense policy.

Patrick Friel is senior consultant to the National Institute for Public Policy, Fairfax, Virginia, and is the author of a number of articles on strategic systems technology and comparative U.S.-USSR strategic programs. He formed Friel and Company, Inc., in 1970, which specialized in strategic missile and space systems analysis. Friel is presently a consultant for ballistic missile defense research and development programs and has previously served in a number of important government positions, including deputy assistant secretary of the army for ballistic missile defense.

Daniel Gouré is director of Soviet Studies, SRS Technologies Inc., Arlington, Virginia. He was previously director of threat analysis, pilot architecture study, Strategic Defense Initiative Organization (SDIO). Dr. Gouré received a Ph.D. from Johns Hopkins University and has contributed to the literature on national security in a number of authoritative studies. Among these are analyses of Soviet countermeasures to SDI, Soviet nuclear strategy and force planning, military conflict in space, and active defenses in Soviet policy and strategy. His current work includes research on trends in Soviet theater defense force structure and planning.

Gary L. Guertner is professor of international relations at California State University, Fullerton. Formerly he held the Henry Stimson chair of political science at the U.S. Army War College and served on the staff of the U.S. Arms Control and Disarmament Agency. Guertner is the author of numerous articles in the professional literature of defense policy and arms control, and he recently coauthored *The Last Frontier: An Analysis of the Strategic Defense Initiative* with Donald M. Snow.

Keith B. Payne is executive vice president and director of national security studies, National Institute for Public Policy, Fairfax, Virginia. Payne has written extensively on arms control issues, Soviet foreign policy and strategic doctrine, and U.S. defense policy. He is the author of *Nuclear Deterrence in U.S.-Soviet Relations* (Westview Press, 1982) and *Strategic Defense: Star Wars in Perspective.* Prior to joining the National Institute for Public Policy, Payne taught at the University of Southern California and was a member of the senior professional staff at Hudson Institute. He is an adjunct professor at Georgetown University.

David N. Schwartz has extensive government experience in strategy and arms control issues and is a leading author in the field of deterrence and Western European defense. Author of *NATO's Nuclear Dilemmas,* Schwartz served in the Bureau of Politico-Military Affairs in the U.S. State Department and in 1985-1986 was on leave from the Department of State at the International Institute for Strategic Studies, London. Schwartz is currently writing about theater ballistic missile defenses for Europe and other issues pertinent to alliance relations.

Leon Sloss is president of Leon Sloss Associates, Washington, D.C., and is a prominent analyst of U.S. national security policy and strategic programs. Sloss directed the nuclear targeting policy review studies for the Carter administration and in that capacity and subsequently has developed some of the more original U.S. policy constructs related to escalation control, flexible targeting, and war termination. He has contributed to numerous studies of U.S. defense policy and strategic programs, including his chapters in Ashton B. Carter and David N. Schwartz, eds., *Ballistic Missile Defense.*

Donald M. Snow is professor of political science and director of international studies at the University of Alabama. During 1985-1986, he was one of the first two secretary of the navy senior research fellows at the U.S. Naval War College, and he previously served as professor of national security affairs at the U.S. Air Command and Staff College. A fellow in the Interuniversity Seminar on Armed Forces and Society and chairman of the Section on Military Studies, International Studies Association, Snow's recent books include *The Last Frontier: An Examination of the Strategic Defense Initiative* with Gary L. Guertner; *National Security: Issues in U.S. Defense Policy* (forthcoming); *The Eagle's Talons: War, Politics and the American Experience* (with Dennis M. Drew, forthcoming); and *Fencers: U.S./USSR Relations Face the 1990s* (forthcoming).

David S. Sorenson is associate professor of political science, Denison University, and senior faculty research associate, Mershon Center, Ohio State University. His publications in the field of defense policy have included studies in modeling the nuclear arms race, superpower defense expenditures, ballistic missile defense for Europe, and the economic and social impacts of defense spending. His contract research has included analyses of Soviet responses to U.S. space-based defense systems. Sorenson also has done substantial research and publication on comparative foreign policy, especially U.S.-German relations, and has developed innovative classroom simulations, some of which have been published. He is chair-elect of the Section on Military Studies, International Studies Association.

Jonathan B. Stein is a legislative assistant to U.S. senator Paul Simon. He was previously a fellow at the Center for Strategic and International Studies, Georgetown University. He authored *The Soviet Bloc, Energy and Western Security* and *From H-Bomb to Star Wars.* Stein received an M.S. in foreign service from Georgetown University School of Foreign Service and a B.A. from Vassar. In 1980 he worked in the Office of International Security Affairs, Office of the Secretary of Defense, on Warsaw Pact issues.

Richard E. Thomas is director of the Center for Strategic Technology, which is part of the Texas A&M University System. He received a Ph.D. in aeronautical and astronautical engineering from the Ohio State University. Thomas began his Soviet studies in 1951 with what is now the Foreign Technology Division of the U.S. Air Force. Since that time he has conducted research dealing with elements of U.S. and foreign space and weapons technologies. His recent writings have focused on Soviet electro-optical and hypersonic ramjet research.

Seymour Weiss is president of Sy Associates, Inc., a Bethesda, Maryland, corporation that provides consulting services on foreign and defense policy issues. He was ambassador to the Bahamas from 1974 to 1976 and has held a variety of important senior positions in the Office of Management and Budget, the Agency for International Development, and the Department of State. At the State Department, Weiss was director of the Office of Politico-Military Policy, director of the Office of Strategic Intelligence and Research, special assistant to the secretary of state, deputy director of the Policy Planning Staff, and, as assistant secretary of state, director of the Bureau of Politico-Military Affairs. In these positions he was instrumental in formulating and executing policy during the Strategic Arms Limitation Talks, the Berlin and Cuban missile crises, and other important issues. He was named chairman of the Defense Policy Board Advisory Committee in 1985 and appointed to the president's Foreign Intelligence Advisory Board after serving as an adviser to presidential candidate Ronald Reagan in 1980.

James C. Wendt has been senior policy analyst, The Rand Corporation, Santa Monica, California, since 1984 and has been at Rand since 1980. He was previously assistant professor of sociology at Columbia University, New York. His publications include Rand studies on: improving the NATO force-planning process; allied security concerns in theater nuclear employment; extended deterrence and the Strategic Defense Initiative; the effect of Soviet ballistic missile defense on British and French strategic forces; and Soviet sortie generation capabilities and vulnerability. Wendt is a fellow of the International Institute for Strategic Studies, London.

For Product Safety Concerns and Information please contact our EU representative GPSR@taylorandfrancis.com
Taylor & Francis Verlag GmbH, Kaufingerstraße 24, 80331 München, Germany

www.ingramcontent.com/pod-product-compliance
Ingram Content Group UK Ltd.
Pitfield, Milton Keynes, MK11 3LW, UK
UKHW022117190926
475702UK00013B/1495
9780367311933